MV

D0392126

Captivating Life

CAPTIVATING *Life*

JOHN C. AVISE

A Naturalist in the Age of Genetics

SMITHSONIAN INSTITUTION PRESS
Washington and London

Copy editor: Tom Ireland
Production editor: Robert A. Poarch
Designer: Janice Wheeler

Library of Congress Cataloging-in-Publication Data
Avise, John C.
Captivating life : a naturalist in the age of genetics / John C. Avise.
p. cm.
ISBN 1-56098-957-2 (alk. paper)
1. Avise, John C. 2. Naturalists—United States—Biography. 3. Geneticists—United States—Biography. I. Title.
QH31.A85 A3 2001
508′.092—dc21
[B] 2001020635

British Library Cataloguing-in-Publication Data is available

Manufactured in the United States of America
08 07 06 05 04 03 02 01 5 4 3 2 1

© The paper used in this publication meets the minimum requirements of the American National Standard for Information Sciences—Permanence of Paper for Printed Library Materials ANSI Z39.48-1984.

Contents

Acknowledgments

Credit for most of the research described in this book should go to the many wonderful students, postdocs, and technicians who have worked in my laboratory over the years: Charles Aquadro, Marty Ball, Biff Bermingham, Brian Bowen, Robert Chapman, Donatella Crosetti, Susan Daniel, Andrew DeWoody, Michael Douglas, Anthony Fiumera, Cecilia Giblin-Davidson, Matt Hare, Glenn Johns, Adam Jones, Steve Karl, Lou Kessler, Trip Lamb, Mark Mackiewicz, Beth McCoy, Joe Neigel, Bill Nelson, Guillermo Ortí, John Patton, Devon Pearse, Brady Porter, Paulette Prinsloo, Paulo Prodöhl, Carol Reeb, Nancy Saunders, Kim Scribner, DeEtte Walker, Kurt Wollenberg, and Lorenzo Zane.

I wish to thank several friends and colleagues—most notably Betty Jean Craige, Geerat Vermeij, and DeEtte Walker—for helpful comments on drafts of the manuscript. Vinny Burke has been an enthusiastic and able editor, as well as a friend. I am grateful to the Pew Foundation for recent funding, which encouraged this book. Most of all, I want to thank my family—Mom, Dad, Joan, and Jennifer.

Captivating Life

Part One

Becoming a Scientist

Some natural historians seem to begin life with an ingrained and abiding curiosity about the biological world. Others may have lost their contact with nature in youth, only to reconnect in later years. In any event, people who lack this fascination with nature usually cannot begin to fathom what so preoccupies we few who do. I was among those naturalists seemingly born to dig fossils, hunt, fish, watch birds, identify plants and insects, and otherwise investigate the "great" outdoors. My childhood adventures in the north woods set the stage for what would later become a scientific career devoted to an unprecedented marriage of genetics and natural history.

1 A Biologist's Life and Times

More than 20,000 generations have come and gone since creatures recognizably human first appeared on Earth. Yet only a small fraction of men and women, all within the last three life spans, have been privy to this remarkable scientific fact. Across the 4,000 centuries since the full emergence of *Homo sapiens,* people have struggled to feed, clothe, and defend themselves, find mates, and raise children, often praying to gods for guidance. Occasionally, new inventions or discoveries proved useful and were incorporated into cultural traditions. However, only rarely have societies had the means or inclination to sponsor scientific research or other creative endeavors merely for the sake of knowledge or artistic expression. My life has been spent in one such place and time.

My hope here is to enthuse others about my two vocational loves—natural history and genetics—and to tell the story of how I came to find myself at the forefront of a scientific revolution that would wed these seemingly different disciplines. My own enchantment with nature began in early childhood, whereas an appreciation of the science of genetics was a cultivated taste acquired gradually over a thirty-year career. The former love comes primarily from the heart, the latter mostly from the mind. The blend has been a powerful elixir that has kept me intoxicated with the joys of biological inquiry.

My story, I hope, reveals as much about the making of a scientist as it does about the workings of science and nature. Mine is the tale of a childhood preoccupation later turned occupation, of how a backwoods introvert of rather ordinary talent came to occupy center stage in an extraordinary scientific revolution that would merge two otherwise disparate worlds—the laboratory science of genetics, and old-fashioned outdoor naturalism. In the tradition of eighteenth- and nineteenth-century naturalists such as William Bartram and Charles Darwin, I have been an ardent observer and interpreter of nature. However, the tools of my trade are not only binoculars and field diaries, but also the modern equipment and detective methods of genetics coupled with the insights of evolutionary biology.

I was born in 1948, early in the American postwar baby boom. My father had recently returned from the European theater to marry his longtime sweetheart and start a family. During my upbringing, Dad was a federal postal inspector, and Mom, who had majored in home economics in college, was a housewife. There was nothing unusual about our family circumstance that might have presaged the highly unconventional career path in natural history and genetics that I was eventually to take.

My life as a latter-day naturalist in academic garb has been an odd amalgam indeed. A typical month might see me digging for gophers in a Georgia pasture and, a few days later, delivering a plenary address to an international meeting of molecular biologists; slogging through a muddy swamp in South Carolina in search of treefrogs and then serving on a white-collar panel in Washington, D.C., on how best to manage federal lands for biodiversity; seining blind fish from pitch-black caves in Mexico and then quickly returning to a modern genetics laboratory to shed new light on an ancient gene duplication within the animals' cells; wading waist deep in frigid waters of an Alaskan estuary to band eider ducks and a week later lecturing to a sea of biology freshmen in a university auditorium; and composing an abstruse scientific treatise on a technical topic such as phylogeography and soon thereafter speaking in lay terms before a Christian faculty forum on the societal relevance of genetics to religion.

I have repeatedly found myself among the early scientific pioneers surveying the boundaries between historically unaffiliated fields: laboratory science and outdoor natural history, molecular genetics and endangered species, microevolution within species and macroevolution across depths of time, and even the often tense interface between evolutionary science and religion. Why I originally reconnoitered such frontiers probably had much to do with my aversion to crowds—I have always been a maverick, harboring a distaste for herd mentalities. When I have stumbled upon valuable scientific nuggets during my career, it is, I suspect, because I have roamed relatively untrodden ground.

Underlying my research has been an appreciation that there is a universality to life on Earth: all creatures share the same type of genetic material (primarily DNA) by virtue of common ancestry in an evolutionary tree with roots nearly four billion years deep. This molecular commonality also means that standardized tools of genetic analysis can be applied to any species. In taking advantage of this fact over the years, my students and I have employed genetic markers to study organisms ranging from sponges, oysters, and fiddler crabs to numerous fishes, frogs, turtles, lizards, birds, and mammals. In more than 230 research publications, we have examined (to varying extents) the genetic features of more than 400 animal species.

Not all of the exuberant diversity of life is evident to a naturalist's naked eye. Through the lens of modern genetic analysis, we have gained fresh insights into the intriguing evolutionary histories or natural lifestyles of a wild cast of biological characters, such as the nine-banded armadillo, a species in which each litter is composed of genetically identical quadruplets; pipefishes and seahorses, in which males become pregnant and give birth to live young; the Amazon molly, a fish that has done away with males entirely; horseshoe crabs, "living fossils" that look just like their ancestors of 150 million years ago; beautiful warbler species that, despite varied plumages and songs, all proved to have evolved within the last few million years; American eels, which migrate from their nurseries in freshwater streams to spawn in a random-mating melee in the Sargasso Sea; and marine turtles, which somehow are drawn to lay eggs on their own natal beaches after decade-long oceanic odysseys that would put the journey of Odysseus to shame.

Such personal exposure to a wide swath of biology is a far cry from how genetic (or most other biological) research traditionally is conducted. Usually, a biologist devotes an entire career to one organism such as the fruit fly or a fish species, learning as much as possible about the creature itself or plumbing the depths of some particular biological phenomenon. Our comparative method extends this approach by focusing also on the genetic or evolutionary outcomes as contrasted among creatures that may be as different as fruit flies and fishes. Traditional and comparative methods are not mutually exclusive, and of course we have employed both. Nonetheless, it is the universality of DNA, the genetic material of life, that has allowed us to sail on voyages of discovery across a broad latitude in the biological world. Furthermore, for a naturalist like me, forever enthralled by the diversity of nature, the comparative method is probably the only approach to genetics that could have kept my rapt attention for a lifetime.

Apart from the joy of scientific discovery itself, I have been amply rewarded during my career. I have received notable scientific honors from an unusual breadth of sources, ranging from a Sloan Award in Molecular Evolution, to a Pew Fellowship in Marine Conservation, to the Brewster Memorial Prize from the American Ornithologists' Union for decade-long contributions to avian biology. I have served on the editorial boards of more than a dozen scientific journals and as president of two leading scientific organizations (the Society for the Study of Evolution, and the American Genetic Association). I was among the youngest biologists ever elected to the National Academy of Sciences and was elected, also at a young age, to the American Academy of Arts and Sciences for creative accomplishments beyond basic science alone.

History will judge whether my eclectic career may have any significant, lasting impact on science, but the journey itself merits telling for what it reveals about the broader state of nature and the workings of science during the latter half of the twentieth century. Just four years before my birth, geneticists first showed conclusively that an obscure biochemical substance known as deoxyribonucleic acid (DNA) was the genetic blueprint of life. I was five years old when James Watson and Francis Crick announced their discovery of DNA's simply elegant double-helical structure. In that same year (1953), Eugene Odum published *Fundamentals of Ecology,* a seminal textbook that launched a new science on the interrelationships between living organisms and their environment.

Now, a half-century later, both DNA and ecology are household terms. Geneticists recently elaborated the precise sequence of all three billion letters in the human genetic code, a necessary step in deciphering the function of the 30,000 or so genes inhabiting each of our cells. Amazing breakthroughs in genetics, medicine, and the other biological sciences are featured almost daily in newspapers and magazines. So too are ecological and environmental disasters of epic proportions. We live in a time of unprecedented scientific and societal transition, and the fate of the planet hangs precariously in the balance. As a global culture, we struggle to comprehend these revolutionary developments and what they might mean for our future, ironically often seeking guidance in religious outlooks that, for better or worse, differ little from those of our Bronze Age past.

As an evolutionary geneticist, conservationist, ecologist, and natural historian, I can't escape the fact that my beloved science rests at ground zero for some of the most challenging and controversial issues of our times. Indeed, I would not want to escape the social responsibility that comes with this appreciation. I hope that this book will explain how I have come to feel the way I do about the marvels of nature and the artistry of scientific research.

2 Ice Lake

The name comes not from the fact that its waters are chilling, but because it was a source of ice for the town of Iron River. A century ago, a large barn stood along the western shore of Ice Lake, today the site of a town park. Each winter, blocks of ice freshly cut from the lake by large saws were taken there by horse-drawn sleighs for storage between layers of sawdust. Throughout the ensuing year, these blocks supplied the home iceboxes of Iron River's 3,000 citizens.

Ice Lake is my physical and spiritual home, to which I return in mind particularly in times of difficulty. It is located in what the settlers and their descendants refer to as "God's Country": the wild, forested territory of the Upper Peninsula of Michigan. This beautiful land, dotted with lakes and rivers, was under many hundreds of feet of ice just 18,000 years ago, when the latest of the great Pleistocene glaciers pushed into the northern half of the continent about as far south as today's Missouri and Ohio Rivers. These and earlier glaciers scoured and carved the region, exposing bedrock here, depositing till there, and gouging huge divots that were to become the Great Lakes.

Our house, originally my grandparents' home, is situated on the north shore of Ice Lake. My great-grandparents, Charles and Christina Johnson, homesteaded 120 acres in the 1880s, and as each of their sons married, he was given a choice of two acres on which to build a home. Oscar, the second son and the first to marry, chose high ground north of the road, in a time when lake frontage was so plentiful it was not valued greatly. Grandma Selma and Grandpa Bill, the oldest son and the next to marry, chose their plot between the lake and a road

The original homestead, with Ice Lake in the background.

that was the main thoroughfare into town for horse buggies, sleighs, and foot travelers. Built in 1915, their house was one of the first in rural Iron County to have indoor plumbing and prewiring for electricity, which Bill anticipated would soon come to the area.

The house was oriented to take full advantage of the road frontage, with the porch and all large windows oriented in that direction. The back had only small windows facing the lake, almost as an afterthought. When my parents inherited the house many years later, they renovated the back with picture windows and a deck to take advantage of what many people in our generation would consider the primary appeal of the property: a wonderful lake view.

This generational shift in our family's orientation toward natural versus artificial landscapes paralleled broader societal attitudes of the times. When the Upper Peninsula was colonized by European (primarily Scandinavian) immigrants in the last century, they found a rocky, forested land similar in aspect to their homelands in Sweden and Finland. This was a land to be conquered, tamed, and utilized. First came the logging boom. Nearly every mature white pine and other straight-trunked tree was cut during the heady days immortalized in the stories of Paul Bunyan and Babe, his giant blue ox. Logging camps sprouted across the land, each the epicenter of a growing clear-cut that eventually merged with others until the entire Upper Peninsula was denuded of its majestic forests. Early black-and-white photos reveal what to our modern eyes seems a bleak, barren landscape. I wonder what the loggers saw: perhaps the beauty of a hard-won victory of civilization over an intimidating wilderness. The thirty-foot-tall Paul Bunyan, who with his gigantic axe could fell 100 trees in one swing, was fiction, but his human counterparts collectively accomplished similar feats in the decades between 1840 and 1900.

Logging continues today as a major economic activity of the Upper Peninsula, but the commercial focus has shifted from the production of lumber for furniture and housing, to pulp. Now, second-growth birches and poplars are felled and shipped to distant mills for processing into newspapers and magazines. Still, I love these scrubby north woods with their fern-strewn, moss-covered floors. They seem natural, as if the harsh climate and rocky landscape could never have had more grandiose designs for its plant life.

To this day, whenever I return home from the "lower 47," I cross a marked transition zone where the land and my emotions change in concert. Heading through Wisconsin, that point is north of Green Bay; in Michigan's Lower Peninsula, it is just beyond the town of Grayling. Thereafter, going north, the cornfields of the Midwest become a fading memory. Now, white trunks of disheveled birches and poplars dominate, contrasting with the dark greens and blues of alders, maples, and spruces. Bogs, the spongy, vegetation-choked remnants of former lakes, make their welcome appearance. All of the trees shorten, as though to duck their heads below a sky that seems suddenly lower, such that on clear days, puffy clouds sail by nearly within reach. The air cools, and the land takes on a distinctive feel and smell. It is as if a cleansing shower begins to wash from me the sweat and stress of the hectic life in the human stew pot farther south. No more does nature subsist merely in isolated pockets, like out-of-element animals in a zoo or in a small patch of urban forest. Here, nature dominates, and it is the small human settlements that strain for economic survival. Much of the wilderness spirit remains.

After the primeval forests were cut, the next great economic boon to the Upper Peninsula was mining. To the north, in the Keweenaw Peninsula, which juts into Lake Superior, copper deposits were discovered. In our area, iron ore was king, and it stained with an ochre hue the river that lent its name to our town. For nearly a century, beginning in the 1880s, large mining operations such as the M. A. Hannah prospered from the toil of miners and their families, who often lived and died in company towns, or "locations." The extracted mineral eventually was transported, often by distinctive ore boats (some of which still ply the Great Lakes), to the insatiable steel factories of Detroit and Gary, Indiana, which supported our nation's industrial growth, war efforts, and a burgeoning automobile industry.

At the mine sites, elevator shafts were sunk below monstrous red towers that even today remain scattered about Iron County as ghostly testimonials to this bygone era. From these vertical shafts, at varying depths underground, horizontal drifts (tunnels) radiated outward. There, miners pick-axed and blasted the rock, slaving away their lives following the veins of red gold. In later years, long after the mines closed, some of these mine shafts and drifts collapsed, causing "cave-ins" that dot the county today. One such cave-in occurred a quarter-mile from our home. We were awakened one summer night by a rumble and the strong smell of sulfur. In the morning light, we discovered a huge pit, some 150 feet deep and 400 feet in diameter, just across the highway.

My Grandpa Bill worked in iron mining for about fifteen years, first as a diamond driller, then as a lathe operator in the machine shop at the Bates Mine. Failing health, most likely from undulent fever, forced him to quit work at the mine. As he slowly recovered, he became the community handyman, a self-taught carpenter, plumber, and electrician, and the county's expert on septic tanks and their installation. I barely remember my grandfather—he died when I was not quite five—but everyone said he was an exceptionally strong and kind man. He and Grandma were proud of the fact that they had homesteaded in God's Country, and they imbued in me at an early age an enduring love of the north woods and of the fierce self-sufficiency of its inhabitants.

Although I viewed it as home, Ice Lake was really only my part-time residence. During my childhood, after school let out each summer, Mom and I would travel the 500 miles to Ice Lake from our primary house in Grand Rapids, in Michigan's populous Lower Peninsula. Dad, who had a demanding career as

On the Studebaker truck with Grandpa Bill.

a postal inspector, would drive up to join us for two weeks at the end of the summer. In the early years, Mom and I traveled north by train and Greyhound bus. Later, we took commercial flights to the nearest airport, in Iron Mountain, about forty-five miles from Iron River, aboard tail-down DC-3 airliners. These propeller-driven planes seemed huge to me at the time, but now would be dwarfed by even small commercial jets. Dad assured us that DC-3s were the safest planes ever built, and we believed him because he had flown in them extensively in Europe during World War II.

On one memorable bus ride across the Upper Peninsula, the driver yelled out "Wolf!" as a large gray animal bounded across the highway. Grandma had told me stories of her own childhood, when wolves were more common. One late winter afternoon, walking the lonely three miles home after school, she and her sisters had an eerie feeling that something was watching them, and once home they reported as much to their father. He could find nothing in the fading light, but next morning discovered paw prints in the snow alongside those of the children. I was told that the wolf pack that followed Grandma had done so out of curiosity: there are no documented wolf attacks on humans in the Upper Peninsula. Within a few years, the few remaining wolves were extirpated from the region. Ironically, recent years have witnessed successful efforts to reintroduce wolves to the Upper Peninsula, rather than exterminate them.

When I was very young, Grandma was still vigorous and independent. She and Mom worked the minifarm, with my able assistance, of course. My job each day was to gather eggs from a dozen hens in the chicken coop down by the lake (avoiding the fearsome roosters, which were nearly my size) and to help Grandma milk the cow and feed it hay from the barn. In addition to the milk itself, we relished the cream skimmed from its surface and the butter that we churned in a drum-sized wooden vat.

Once or twice each summer, I helped find, dig, and trim small spruce and pine seedlings in the nearby woods for transplantation to Grandma's tree farm, which brought in modest proceeds each Christmas. It also fell (in small part) upon my skinny shoulders to help plant and later to harvest vegetables from an unkempt garden that sprawled across more than an acre. We grew everything, it seemed, in the rocky soil: red and white potatoes, carrots, celery, beets, onions, peas, green and yellow beans, rutabagas, cabbage, brussels sprouts, turnips, lettuce, rhubarb, squash, green peppers, tomatoes (in good summers), cucumbers, and all manner of herbs such as parsley and basil.

Strange as it seems, soil in the Upper Peninsula also grows rocks, typically about three to ten inches in diameter. These appear anew each spring when the dirt is tilled, having been heaved up from deeper soils by cycles of freeze and thaw. They are another signature of glacial influence, bulldozed by sheets of ice

Digging potatoes with Grandma Selma.

and deposited in abundance on the northlands over the last two million years. Nearly every small farm today is dotted with rock piles and bordered by stony palisades. The hopelessly rocky soil is one reason why the Upper Peninsula was never converted to the wheat fields of the plains states to the west or the vast expanses of corn and soybeans to the south.

The land surrounding our Ice Lake home provided even more bounty. In mid- and late summer, wild fruits abounded: crabapples, raspberries, blackberries, pincherries, gooseberries, and chokecherries. Wild blueberries grew on the Lake Mary Plains, a sandy area of bracken ferns and scattered jack pines about twenty miles away, near the town of Crystal Falls. We drove there in our 1950 Studebaker pickup, a bluish-green truck whose starter switch was under the accelerator pedal and whose windshield wipers moved at a variable rate, in harmony with the truck's forward speed, which never exceeded 40 mph. I rode on blankets and cushions in the back. I can still picture the bed of that pickup in perfect detail: the ribbed floor, the rounded wheel wells, the two rubber-covered chains whose terminal hooks secured the tailgate, and the plaintive squeaking of the tailgate on its hinges. It's funny how the small things sometimes are remembered.

The Lake Mary Plains was the only place in the Upper Peninsula where I recall feeling uncomfortably hot. Under the sizzling sun, with two-quart pails secured around our waists by rag belts, and with my blue transistor radio tuned to a Detroit Tigers baseball game, we roamed the area for hours. Grandma was the fastest picker, using a raking style that brought into the pail as many leaves and twigs as it did blueberries. Mom and I picked clean and filled our pails more slowly, but claimed that we saved time in the long run because we did not have to remove debris later.

To understand our harvesting compulsion, it is necessary to appreciate the Swedish psyche. In the Upper Peninsula, as in Scandinavia, winters are long and harsh, the summers brief. The growing season lasts a scant 100 days, and no month is guaranteed to be free of frost. The impulse to set stores for the winter came as naturally to us as it did to the region's chipmunks, red squirrels, and black bears. In our case, this meant canning fruits and vegetables, making preserves, and laying in the necessary cords of firewood to last until the following spring.

By late summer, Grandma's kitchen was rich with the sounds and smells of the harvest being processed. Boiling slurries of sliced fruits and vegetables bubbled in metal vats atop the giant wood stove, soon to be poured into canning jars that accumulated by the score. Beautiful jams and jellies from the apples and wild berries were sealed in smaller jars under a layer of paraffin-wax. Pies for immediate consumption were always available too. My favorite was rhubarb pie, sweet and tart, which went well with home-baked Swedish rolls and limpa bread.

I was too young to help with the woodcutting, nor did I have much taste for watching Grandma use the stump out back as a chopping block for dispatching an occasional rooster for dinner. I was quite intrigued, however, that the decapitated bodies were able to run spasmodically around the yard for several seconds after the bloody deed!

I was proud that I could help in other ways. My most enjoyable task was to supply fresh fish. Nearly every day, weather permitting, I rowed our small boat to one of several fishing holes on Ice Lake. These were specific drop-offs a few hundred feet from shore where the bottom vegetation first sank away to bare visibility under eight feet of water. There, with hook baited with worms freshly dug from the garden or inch-long crayfish captured among shoreline rocks, I dropped my cane-pole line to catch yellow perch, bluegill, bright pumpkinseed sunfish, and wily smallmouth bass. On calm days, I confess to having spent much of my time peering over the side of the boat, captivated by viewing the schools of fish below the glasslike surface.

Often I caught so many fish that Mom and Grandma encouraged me to sell or donate much of my catch to neighbors and town folk. It is probably fortunate that we maintained a varied diet. Today, Michigan's Department of Natural Resources warns against eating fish more than once a week from any of the state's waters, in part because of mercury contamination. This toxic metal was

Displaying a fresh catch of fish from the lake.

used indiscriminately to process iron and copper ores during the heyday of the mining era.

When Dad arrived late in the summer, he would often arise before daybreak to troll Ice Lake for its northern and walleye pike. By the time I awoke, he had returned to the house, often with a stringer of fish, which he left hanging on the knob of the pantry door for all to see. As I got older, I accompanied Dad and my Uncle Elmer on fishing trips on Ice Lake and to dozens of other lakes and streams in the area.

Each body of water had a distinctive biological signature. Iron Lake was good for slab-sized crappie. Lake Ottawa was known for rock bass and larger sunfish, and also was home to a nesting family of bald eagles. Pine Lake and Stanley Lake had muskies, pikelike predatory fish up to six feet long that always made me think twice before swimming there. Golden and Hannah-Webb Lakes, spring fed and crystal clear, were stocked with rainbow trout. Many of the smaller creeks had native brook trout. The Paint River housed smallmouth bass, caught by casting worms from shore or lures from a drifting canoe. Iron Lake was noted for large northern pike, as was Perch Lake, despite its name. Chicaugon and Fortune Lakes had lovely sand beaches and were nice places to swim and catch small panfish. Lake Emily was home to largemouth bass and walleyes. One of my most notable fishing experiences with Dad was netting a twenty-eight-inch walleye after a battle that seemed to last an eternity. Lake

Emily also was known for its jumbo perch. When they were biting, word spread among the locals, and the lake soon would be dotted with rowboats projecting cane poles, giving each boat a prickly appearance, like one of the many porcupines that roamed the adjoining woods.

On occasional longer excursions, we visited Lake Superior, about ninety miles to the west, and trolled its mighty waters for lake trout. When I was little, Dad and Elmer would leave me with Mom, Grandma, and Aunt Alice to hunt for agates along the rocky shores and to swim (ten minutes maximum) in the lake's icy waters. When I was older, Mom and I joined the fishing crew. We always kept a close eye on the weather, because ferocious storms from Canada often swept across Lake Superior with no advance warning beyond a black line of clouds on the horizon. After beating a retreat to port and finding safe vantage, we would watch in awe as waves built to great crashing surges against the rocky cliffs.

We made it a point to visit a different lake every week or two of the summer, yet after more than a dozen years, we had only begun to canvass the waters of Iron County alone. One great appeal to me of the Upper Peninsula is that the land is sparsely populated yet accessible. Paved and gravel roads, many the legacy of timber and mining days, crisscross the region and open it to casual modern-day exploration. This infrastructure was bolstered during the Great Depression when the federal government employed thousands of workmen in its WPA (Work Projects Administration) and CCC (Civilian Conservation Corps) camps to build roads and parks in the region. Now, nearly every lake has a public access with boat ramp and campsite maintained by the state, county, or township. It is not unusual to find oneself alone on the shores of a splendid lake or river that in the lower 47 would be swarming with visitors. The first roadside park in the United States (according to the sign) is still present, just a few miles east of Ice Lake.

The Upper Peninsula is one of the few regions of the country where the human population declined in the latter half of the twentieth century. Once stripped of its evident natural resources—primary forests and ore—the area's economic base collapsed, and younger generations left to find employment in urban areas to the south. During all of my time in the U.P., I had virtually no contact with other children. There were a few kids in town, to be sure, but none lived nearby. That didn't matter—I was fully engaged in outdoor activities and found ample companionship in Ice Lake's wildlife.

As I grew, I became quite the young entrepreneur. I sold worms and nightcrawlers by the dozen to fishermen, advertising my business on a hand-painted sign in our front yard. Small worms were dug from the garden or coaxed to emerge from the lawn by electrodes from a small, hand-held generator. Night crawlers were gathered at night from the community golf course, where eight-inch beauties lay sprawled about in abundance. I stored them in a barrel filled with sphagnum moss that I had dug into the ground in our backyard. The worms required little maintenance: just a small pinch of corn meal every day or two.

Grandma was a skilled weaver and taught me the trade. An upper bedroom housed two full-sized looms and mountains of balled rags. The extended Johnson clan routinely sent Grandma retired clothes and old sheets. She and Mom would cut these into inch-wide strips, sew them end-to-end, and roll them into balls of every color and texture. These provided the raw materials to be crafted into rugs or place mats, sometimes churned out at a rate of two or three per day. Once the warp is strung on the loom, the process is simple: spool the rag strips onto a wooden shuttle, pass the shuttle through threads of warp spread in elaborate designs by attachments to foot treadles, and pound each successive strip of weft into place using the comblike beater. The harder the pounding, the tighter the weave. I had strong arms, so my rugs were taut and durable. In addition to decorating our home with these rugs and wall hangings, I draped many of them on lines strung in our front yard for sale to passing motorists.

Another source of entertainment and income for our family was the annual Iron County Fair, held in late August. We would load the Studebaker with entries to be transported to the fairgrounds: canned goods, jams and jellies, woven rugs, place mats, hand-sewn clothing, a garden exhibit of fresh vegetables, and various other displays such as a rock-and-mineral collection that I had assembled during trips to local mine dumps and quarries. For a wildflower arrangement, on the way to the fairgrounds we stopped to pick a few of the many daisies, goldenrods, black-eyed Susans, forget-me-nots, Queen Anne's laces, and tansies that brighten the Upper Peninsula in late summer.

The county fair showcased a wonderful menagerie of north woods farm life. In a large circular pavilion, local 4-H kids paraded their groomed cows and heifers in front of judges and an approving audience of parents. In converted stables were housed prized chickens, rabbits, sheep, pigs, goats, horses, and other farm animals, as well as canning and weaving displays, and other arts and crafts. On judging day, I scurried between buildings monitoring our entries for blue, red, and white ribbons, signifying first, second, and third place. The fair's catalog listed the monetary reward for each ribbon. One year, we totaled nearly $120 in winnings, a substantial sum of money then.

August was a bittersweet time because it signaled my impending return to school life in the Lower Peninsula. The final few days were ritualized: the last fishing trip; the last swim in Ice Lake; the final family drive along Pentoga Trail, where the forest arched overhead like a cathedral; the last glimpses of eagles and loons; the last harvest from the garden; and good-byes to Grandma. The morning of departure was particularly sad. In glum slow motion, I helped load the trunk with canning jars, which Dad feared would ruin the shock absorbers of his tail-finned Plymouth. As we drove away, I faced the reality that I would not see my cherished north woods for another interminable ten months. Always, I cried quietly to myself as Ice Lake disappeared from view behind the car.

3 Budding Natural Historian

Recollections of our youth provide only a limited and clouded window to the past. Yet that blurred vision, whether faithfully reflecting reality or constrained and subjectively warped, remains instructive as a partial record of who we are or, at least, perceive ourselves to be. I have reminisced on my childhood images periodically over the years and in many cases can no longer say which are factual historical accounts and which are merely distorted memories of memories, modified from the original like a long narrative passed along a chain of storytellers. Nevertheless, these feelings have been influential in my life.

We lived on Paris Avenue, near the city center of Grand Rapids, Michigan. From kindergarten through second grade, I walked every day to Vandenberg Elementary, about a mile away, often with my friends Mary Lou and Ned. I retain only scattered images of those days: the tattered vinyl sleeping mats that we jostled over at nap time; being terrified at recess to approach the basketball area, where the big kids played; shooting marbles in the dirt of the schoolyard's inner playground; and scrambling for access to the prized painting easel in Miss Withey's kindergarten classroom.

I was about five when Mary Lou and Ned moved away, their families being among the first to join the "White flight" from the inner city. I have clearer recollections of times with my new Black friends and of a childhood in what had become a predominantly African American, lower-middle-class neighborhood. The lot of us then attended Henry School from the third through sixth grades. We were straight out of the *Fat Albert* cartoon show by Bill Cosby. There was

Gaylord (Skunky), whom we hesitated to wrestle due to his pungent body odor; Lonnie, a born leader; Bruce, an all-around athlete; Frannie, a freckle-faced bully who was one of the few other White kids in Henry school; my best friends Willie and Junior; and a cast of others.

I was the scrawniest of the bunch, but I had a valuable asset: speed. Being one of the two fastest kids in Henry School, I was always chosen early in team pickup games of dodgeball, football, and basketball. Indeed, I enjoyed a rather protected status: tougher friends and teammates often came to my physical defense in sporting events that sometimes got a little out of control. Morning and afternoon recesses were our favorite part of school life. We would burst from the confines of the classroom to a gravel play yard, bordered by a chain fence topped with barbed wire, to play for a frantic twenty minutes before the dreaded bell demanded our return to class. It was a rough-and-tumble life, as evidenced by my routine of breaking some bone (in the finger, wrist, elbow, or shoulder) every other year. It seems I was always in a cast or in rehabilitation from some such injury.

Our home on Paris Avenue was a two-story brick and stucco, where Grandma Edna lived on the upper floor and our three-member nuclear family below. Edna, blind and nearly deaf for the more than sixteen years that she lived with us before her death, was a stern Christian Scientist. From conversations overheard, I had come to believe that her physical disabilities might have been prevented had she agreed to see a doctor before her vision and hearing deteriorated irreversibly. But she had resisted medical care at critical times, believing that the only path toward healing and salvation was to place her fate in the hands of God. Not only was Edna horribly debilitated for most of her adult life, but she also seemed to blame herself for a lack of adequate faith in a loving God who otherwise would have cured her. All of this registered in my young eyes as compelling grounds for suspicion of devout religious beliefs.

Many evenings, especially when my parents left me in her care for brief vacations, I read aloud to Edna from books or magazines. Many of these were from a condensed-book series published by *Reader's Digest.* I was also asked to read extensively from Mary Baker Eddy's writings and from the Old and New Testaments, whose book titles I was obliged to memorize and recite in order each bedtime.

Despite Edna's physical handicaps, I felt little pity for her at the time. In contrast to my nonreligious (but spiritual) Grandma Selma, who was down-to-earth and loving, Edna maintained an aristocratic, holier-than-thou attitude, with attention focused primarily on the hereafter rather than on family and friends. This was another aspect of her stern orientation that I came to despise and in general to associate with religious fundamentalism.

We had a small family. The only other close relatives were Mom's brother Dick, his wife Helen, and their two daughters. We visited the cousins regularly on major holidays. They lived in the farm town of St. Johns, a ninety-minute drive away. Uncle Dick owned a sawmill and was a true outdoorsman. I loved to accompany him on walks through a large woodlot behind his house. In the spring, he scarred each maple tree with a V-shaped abrasion and tapped it with a spigot, from which he hung an aluminum pail. We made the rounds, gathering pails of sap and carting them to an open shed, where hot coals burned beneath a flat-based metal vat. There, the sticky concoction was stirred and boiled down to maple syrup, destined for hand-labeled jars that we distributed to family and sold to the public.

Our home on Paris Avenue, near the inner city, offered few wildlife experiences. However, when I was very young my parents gave me a copy of Peterson's *Field Guide to the Birds,* and I pored over its illustrations for hours on end. A case can be made that Roger Tory Peterson was the most influential conservationist of the twentieth century, via his pathbreaking concept of the pocket field guide. During his lifetime, Peterson and collaborators produced a wonderful series of identification manuals to various animal groups, trees, wildflowers, rocks and minerals, and other natural-history subjects. These aids to field identification opened the natural world to the public in an organized, accessible manner for the first time. They inspired untold numbers of budding naturalists, myself included.

At the age of seven or eight, with book and the family binoculars in hand, I made my first birding expedition down the gravel alley behind our home. On the official checklist at the front of the book, I dutifully ticked off house sparrow, starling, rock dove, mourning dove, chimney swift, and common nighthawk as these species made their appearance. The list expanded greatly on trips to my cousins' house and on summer vacations to Ice Lake and elsewhere.

I recall the joy of discovering a nesting pair of chestnut-sided warblers, a species that still occupies a special place in my heart. A related creature for which I have fond memories is the endangered Kirtland's warbler, currently numbering only about 2,000 individuals total. This habitat specialist breeds solely in the open jack pine forest in the north-central part of Michigan's Lower Peninsula. On one family trip to the area, I located a handsome male singing near his nest. I always thought that Kirtland's warbler (rather than American robin) should have been designated Michigan's state bird, because it alone among the Earth's 10,000 avian species chose our state as its sole breeding home.

My way of learning birds was to memorize identification cues in the field guide so thoroughly as to be able to recognize, at first sight, almost any newly

The author and his parents at the evening tally following a Christmas bird count of the Grand Rapids Audubon Society.

found species. My parents never ceased to be amazed at the success of this method. For example, on one trip to Dad's home in Mason City, Iowa, I often called out "stop" or "back up the car" so we could confirm my premiere sighting of a bobolink, dickcissel, or yellow-headed blackbird. Perhaps inspired by my aptitude and involvement, Dad himself eventually became president of the Grand Rapids Audubon Society, although even he would admit that his primary qualifications for that job were organizational skills and social acumen rather than birding expertise. In any event, I still follow the same protocol today. Before departing on any business trip to a foreign land, I purchase and study the relevant bird guides so as not to be overwhelmed when encountering new species.

My youthful interests in nature extended well beyond birds. At one time or another I collected nearly everything collectable and identified nearly every article of nature identifiable from a field guide. I developed a large collection of rocks, minerals, and fossils, all cataloged and neatly stored in wooden chests. One major boon to this endeavor was my discovery of an extensive rock outcropping that housed plumose mica, a special form of translucent muscovite that had been laid down in fanlike plumes against a granitic base. With my rock hammer, I hacked out many of these mica specimens from my secret location and, whenever an opportunity presented itself, offered them in trade at rock shops.

I was particularly fond of fossils. On one family camping trip through the western United States, I found ammonites and baculites, long-extinct forms of mollusks with coiled and straight shells, respectively. Lower Michigan itself had shale quarries housing fossil trilobites, crinoids, corals, brachiopods, and other creatures from shallow seas that covered the area more than 250 million years ago. On one class outing to a nearby rock quarry, my goal was to find imprints of *Lingula,* a brachiopod superficially similar in appearance to a small, stemmed clam. I struck out, but one of my classmates found a specimen that he happily traded for an impressive but commoner trilobite from my collection.

Among my other collections were Polaroid photographs and pressed specimens of wildflowers, and a mounted series of butterflies and moths. I was forever begging Mom to drive me outside the city to fields and woods, where I could chase down butterflies with my insect net, a homemade device of thin cloth sewn into a basket and strung around a clothes hanger. I would catch a regal fritillary, red-spotted purple, or spicebush swallowtail, dispatch the specimen in a fumigation jar, and later secure its wings against a drying board. I took care to arrange each mounted butterfly correctly (as described in books), with the base of its forewings extended perpendicularly rather than obliquely from the body. Insect display cases adorned the walls of my bedroom. Together with nature photo albums and field guides in the bookcases, and the rock and mineral collections in chests of drawers, these gave my living quarters the cluttered look of a natural history museum.

One of my best insect finds occurred right on Paris Avenue. While I was playing basketball one afternoon, a big moth flew overhead and landed on the balcony of a nearby house. I quickly fetched my net, and was rewarded with a spectacular polyphemus moth, one of the largest and most beautiful species in the region. I can still picture the deep purple eyespots against the delicate cream-brown of the hindwings, the velvety texture of the moth's wings and body, and its robust, feathery antennae.

One cherished family ritual took place on two or three Saturdays each spring and fall. Raring to go after a restless night of anticipation, I would awaken Mom and Dad at 4 A.M. and herd them to the car for the drive to Lake Michigan, forty miles away. At first daylight, the ramshackle fishing camps at Grand Haven or Holland were places of palpable excitement. There were bait minnows to be purchased from holding tanks beside the dock, cane poles to be scrutinized before renting one of perfect length and feel, orange life vests to be strapped on, and proper seating positions to be selected aboard a twenty-person dinghy whose operator stood amidships at the controls. Hope ran high as we cruised out of the harbor, peering through the fog to catch the first glimpse of breakwaters that would tell us of the wave conditions on the offshore piers.

Like fragile barrier islands, these piers rose a scant six feet above Lake Michigan's often turbulent surface. There, I was in my glory! With a mournful foghorn providing the background musical score, I orchestrated the proceedings, scouting the half mile of cement and rocks for the latest fishing hot spot, baiting hooks, and hauling in yellow perch, gradually cramming our five-gallon buckets to the brim. The legal limit was fifty fish per person, and Mom, Dad, and I often caught our quota. We returned home by mid-afternoon to fillet our extensive catch.

To naturalists young and old, one of Grand Rapid's greatest resources was its Public Museum. This magnificent old building housed wildlife exhibits, mineral displays, and educational expositions on topics ranging from the deep evolutionary history of life on Earth to the recent social customs of Native Americans and early European settlers in the area. The museum was a focal point for everyone in the community interested in nature, human history, and environmental issues.

Upon entering the museum, an unavoidable sight was the skeleton of a gigantic blue whale hanging in the central foyer. To the right were exhibition corridors displaying beautiful fossils as well as mounts of local animals in lifelike poses. Upstairs, in halls dedicated to recent human history in the region, exhibits described in loving detail the lives of native Potawatomis, Ottawas, Chippewas, and recent European colonists. My favorite exhibit, however, was along a darkened concourse on the lower floor, where backlighted dioramas kept me spellbound: a pack of gray wolves stalking a moose in deep snow; a lily-covered pond with a cutaway of the lodge and dam of a family of beavers; a school of perch and trout along a Great Lake's shoreline; and white-tailed deer in an autumn woodland. With nose pressed to glass, I was mesmerized by these compelling scenes.

Several years ago, I returned to Grand Rapids for my thirty-year high-school reunion, and one of my main yearnings was to see the old Public Museum once again. I was deeply saddened to learn that it is no longer in existence, having been moved and transfigured into a glitzy new museum devoted mostly to the history of the furniture industry, for which Grand Rapids is famous. The modern museum complex does retain a few shabby remnants of the original natural history displays that so captivated me as a child, but these have fallen into disrepair and are squeezed into a tiny area with the distinct aura of a curio shop. It is noteworthy that concepts of evolution no longer are mentioned, much less elaborated, in the dimensionless descriptions that now accompany these exhibits.

When I was eleven, the news that I had been dreading finally came: we too would soon be moving to the other side of town. It's not that I loved Paris Avenue particularly (certainly not in comparison to Ice Lake), but I did feel comfortable with my friends and with a distinctive way of athletic alley life that now would be left behind. Despite my skin color, I never felt out of place at Henry School. My contemporaries had accepted me into their community, at least as far as I could tell. If I felt social unease, it was on rare occasions when I was in the exclusive company of White people! Now, I was to be immersed forever in that intimidating, foreign world.

My transition to all-White Ridgeview Junior High was extremely difficult. I felt ostracized and far out of my element. Many of the kids already knew one another from grade school and belonged to tight cliques. They seemed self-assured, brash, and competent in a communal early-teen mode that under any circumstance would have been foreign to my prior experience.

Several ingredients in my early childhood probably contributed to my disconnected if not disenfranchised view of societal norms. Being an only child surely helped set the stage, as did my summers at Ice Lake, structured around the farm and nature. In Grand Rapids, growing up in a Black neighborhood did little to give me a sense of identity with any cultural or ethnic group that realistically could be retained for life. Now, at Ridgeview, I felt distanced from the action, almost as if I was a disguised alien plunked down in the midst of human affairs. Eventually, I did integrate into Ridgeview society and adopt the usual interests in teen social life, but an atypical cultural background may be at the root of my continuing jaundiced view of many societal conventions.

These feelings of disengagement were abetted, I suspect, by my early exposures to fundamentalist religions. For reasons mentioned earlier, I had gained a personal distrust of ardent Christian Science at a young age. Now, in our new home near Ridgeview, we were surrounded by members of the Christian Reformed Church, another fervent biblical lot.

My parents attended Fountain Street Church, an isolated bastion of liberal thought in otherwise conservative Grand Rapids. Ministered by the brilliant Duncan Littlefair (sometimes referred to as Drunken Littlefaith by local townspeople), Fountain Street was an annoying pain in the side to the Christian Right. Dr. Littlefair's sermons were provocative dialectics, often taking unpopular stances on current affairs such as the Korean War, or throwing out intellectual challenges such as understanding inhumanity, or how various cultures develop notions of sin and morality. Our neighbors knew well of my family's affiliation with this devilish house of worship and, with all seriousness and concern, occasionally informed us that we were headed straight for hell.

Such comments bothered me little, at least outwardly. By age thirteen, rightly or wrongly, I had moved beyond what I interpreted to be simplistic notions of God promoted by fundamentalist Christian doctrine. I was much more interested in thinking about how a god could have constructed the cornucopia of nature, with all of its beauty and gruesomeness, its order yet chaos. If a god were to be understood through routes other than blind adherence to spiritual dogma, the search would have to extend well beyond the confining walls of standard religious temples.

Our new home was about a mile from Ridgeview Junior High in one direction, and Ottawa Hills High School in the other. During these years, three exceptional summer experiences had a profound impact on my growing interests in nature and my eventual choice of a career path.

The first of these was an opportunity that arose following the seventh grade. Our science teacher made arrangements to take several of his top young scholars on an "educational" sailing cruise through the Bahama Islands. I was among those invited, and the price was affordable. We were "Gulliver's Travelers," named for the eighty-foot schooner, *Gulliver,* which would be our seafaring home for two incredible weeks. With ten wide-eyed kids filling two station wagons, the courageous teacher and his wife drove us to Miami, and we flew to Nassau to board the ship.

Gulliver was a magnificent two-masted schooner with teakwood decks, a galley with a full-time cook, and bunks and living quarters more appropriate for nobility than a gaggle of barefoot kids. As we cruised the outer islands, sails billowing in the wind, I often rode in the rigged netting alongside the bowsprit. Porpoises sometimes cavorted just below, and flying fishes dodged to the left and right. Each evening found us anchored in a different harbor, going ashore to visit a local town, or at other times content just to snorkel in the crystalline waters.

It was on this trip that I first became enamored with undersea life. The reefs and sandy banks of the Bahamas were alive with extraordinary creatures such as I had never seen before—hard and soft corals, sea anemones, jellyfish, sponges, bristle worms, queen conchs and other mollusks, brightly colored crabs and shrimps, brittlestars, starfish, and sea cucumbers. Fish were abundant, too: elongated trumpetfishes; beautiful angelfishes, whose adults and young differ so greatly in color; goatfishes, which in roving herds probe the sand with their chin barbels; pert damselfishes defending their algal gardens; and fearsome-looking barracudas, which hovered close by as we snorkeled about.

The Bahamian trip was great fun but also served well its educational function. One small lesson came the hard way. Abundant on many reefs were beautiful purple and yellow sea fans, about the size and shape of a small palm frond. I couldn't resist packing two specimens in my luggage as souvenirs. Despite their plantlike appearance, these gorgonians are colonial animals related to corals in the phylum Cnidaria. I was generally aware of this at the time but didn't consider the ramifications. When I opened my suitcase back in Grand Rapids, I was nearly bowled over by the smell of rotting protoplasm!

The trip made a lasting impression. I was enraptured by the diversity and beauty of the ocean. Especially when underwater, I felt like I was in biological heaven, and from that point on knew that I wanted to learn far more about the sea and return to it often.

The next such opportunity arose several years later. Long before the era of accomplished nature programming, so evident on television now, a major source of vicarious natural-history adventure was provided by visiting filmmaker-lecturers. In Grand Rapids, the Public Museum and Audubon Society sponsored about a dozen travelogues each year. In a local school auditorium, explorers from around the world narrated "home" movies of their natural history experiences in distant lands. The topics ranged from camping among moose and bears in Alaska, to exploring the headwaters of the Amazon. My family religiously attended these presentations, and even though they were quaint by today's professional filmmaking standards, I remember them with great fondness.

One such movie was narrated by Harry Pederson, who for years had spent his summers filming undersea life in the Bahamas. I sat spellbound during Mr. Pederson's presentation and could hardly wait to meet him afterwards. As one of the official host families for visiting lecturers, we had been in communication with Mr. Pederson by mail and learned of his need for a diving assistant in the Bahamas in the coming year. Our conversation that evening eventually led to a job offer, which I accepted in about one millisecond.

That summer was another fantasy come true. Mr. Pederson and his family rented a modest thatched-roof cottage between multimillionaire homes in the exclusive Lyford Cay development outside of Nassau. (The swimming pool of one nearby mansion was the site of the famous shark scene in the James Bond classic, *Thunderball.*) Our little home, nestled among trees, was situated beside a sandy beach and boat dock, about a half mile from our underwater filming site in a sheltered bay. The setting was idyllic in other ways too. We were often invited by curious neighbors to elegant cocktail parties, where I would proudly relate our diving exploits as if I were a brave young Jacques Cousteau.

Mr. Pederson taught me SCUBA diving the old-fashioned way. He walked to the end of the dock, dropped the tank, regulator, wetsuit top, weight belt, fins, and mask to the bottom in about ten feet of water, and told me to swim down and don the gear before returning to the surface. It's actually much easier than it sounds if you remember the initial tricks (which he graciously mentioned): turn on the air, place the regulator in your mouth, and breath through the mouthpiece. The rest can be done at one's leisure, awkward though it may be. I took to SCUBA like a fish and have been diving ever since.

Our primary mission that summer was to film an extended sequence on the jawfish, a curious little creature that lives in self-excavated burrows under fifteen feet of water in protected bays. This sequence would be incorporated into Mr. Pederson's films for his lecture circuit the following winter. Mr. Pederson and his son Chris did the actual filming, using movie cameras housed in homemade casings that had to be air-pressurized each day with a bicycle-tire pump to keep water from leaking in. For stability during the filming process, Mr. Pederson and Chris remained anchored to the sea floor by lead boots and weights. They received air via hoses connected to a compressor in our little boat above. By contrast, I was SCUBA equipped and mobile. My jobs were to deploy the translucent screens that served as light filters under the intense sun, gather particular forms of marine life that the Pedersons wanted filmed, and ensure that the compressor kept running.

Very few nature films existed in those times. I quickly learned that filmmakers often concocted artificial situations for entertainment purposes and gave anthropocentric interpretations to the animal's behaviors. Even the best of the early nature classics, such as Walt Disney's *The Living Desert* (1953), appear contrived by today's standards, and Pollyanna-like in places. It was in that filmmaking tradition that we set about to portray a typical day in the life of a jawfish.

Our hero was an engaging creature painstakingly caught in various behavioral acts over several weeks of filming effort. With gaping mouth out of proportion to his four-inch body, the macho jawfish was filmed excavating his burrow, lining its entrance with pebbles, lounging on his doorstep, feeding on brief excursions from home, and threatening territorial neighbors. We played a few dirty tricks on him, too. In one sequence, we lured him out and quickly slid a glass plate over his tunnel entrance, much to the protagonist's puzzlement and consternation. In another sequence, we slipped a mantis shrimp into his burrow. The shrimp is an excavator, too, and has powerful raptorial claws that can deliver a sharp pinch. You can imagine the look of surprise on the homely face of our jawfish upon his tail-first return to his shrimp-occupied home!

During this magical summer, we also encountered and sometimes filmed many other fascinating creatures. One day, a majestic manta ray glided into our bay, wings gently flapping. Chris and I quickly rigged a tow rope to our motorboat and took turns trolling ourselves through the water just above the seven-foot creature, so close that we could touch its back. On another occasion, I netted a

rare indigo hamlet, a five-inch fish with vertical purple-blue stripes. Having never encountered this species before, I took it to the Nassau aquarium, where the curator put it in a display tank and then gave me a personal tour of the facility.

Like the first, my second summer in the Bahamas transformed me. I was again captivated by the biological world and saw ever more clearly that I wanted to devote a lifetime to its study. I was also introduced to the concept of observational natural history as a scientific enterprise in its own right, in which novel secrets of nature might be learned with suitable diligence. One such disclosure by Mr. Pederson impressed me deeply. A few years earlier, he had discovered a blue-and-white-striped shrimp previously unknown to science. Usually associated with sea anemones, this lovely species sets up cleaning stations frequented by larger fish, who seek the shrimp's services in removing external parasites. Museum systematists gave this cleaning shrimp the name *Periclimenes pedersoni,* in recognition of its discoverer. I thought that this must be the highest type of honor to which any biologist might aspire.

A third summer experience, made possible by the National Audubon Society, followed my senior year in high school and further spurred my interests in natural history. Then and now, the Audubon Society maintains instructional camps around the country where teachers or other adults can receive natural-history training during intense two-week sessions throughout the summer. I applied to work as a junior counselor at one of these facilities and in 1966 was offered a position at the Sarona Camp, in the northern woods of Wisconsin.

This facility consisted of a single-story dormitory building with wings for visiting campers; cottages for ten faculty; another small dormitory for the staff, including three of us junior counselors; and a large renovated barn, where meals were served and indoor classes offered. The campground also had a pavilion for evening lectures and amenities such as a volleyball court, nature trails, and swimming dock on an adjoining lake.

The title of junior counselor was a bit ostentatious. In exchange for room and board, our main job was to wash piles of dishes. Three times a day, we performed these and related bussing services for the several dozen campers who attended each session. Other odd jobs were done as well, at the director's bidding. For example, he told us surreptitiously to rid the camp of feral house cats, which wreaked havoc on the native wildlife.

The job's major perk was that it allowed us voluntary participation, time permitting, in the course offerings. Each class consisted of a blend of lectures and demonstrations by the professional staff, hands-on workshops, and field trips. For example, the aquatic ecology group used the camp's pontoon boat, *Potamogeton* (named after a pondweed), to cruise a nearby chain of lakes and bogs in quest of micro- and macroscopic pond life, often for further examination under a microscope. The geology group visited rock quarries and interesting landforms, and the entomology, ornithology, and botany groups made similar expeditions. Over the four or five summer sessions, I managed to partake of nearly all of these educational opportunities.

In high school I also became increasingly concerned about human overpopulation and its direct and indirect consequences for the Earth's biota. Indeed, I was often preoccupied by such thoughts. One term paper project in particular influenced me deeply. In newspaper accounts, I had read about extensive avian die-offs on the campus of Michigan State University, for example. I arranged to interview an obliging professor there, who had been working on the problem. He told me of hundreds of American robins recently found convulsed on the campus, and even as we spoke, another spasmodic, dying bird was brought into his office by a graduate student. Poisoning by DDT was suspected.

Before the use of DDT was banned in the United States, this synthetic pesticide (highlighted in Rachel Carson's 1962 book, *Silent Spring*) was sprayed indiscriminately in efforts to control insect pests. Unfortunately, beneficial insects suffered too, as did many avian species. Peregrine falcons, brown pelicans, and some other birds near the top of the food chain were nearly extirpated from the continent when long-lasting DDT residues caused eggshell thinning and reproductive failure. Peregrines and brown pelicans later recovered thanks to concerted efforts by environmentalists, but the specter of extinction for these and other high-profile avian species had loomed dangerously close.

In the Michigan State case, DDT had been deployed in a mostly futile effort to save the campus's stately trees from Dutch elm disease. This disease is caused by a fungus inadvertently introduced to North America in a shipment of logs from Europe earlier in the century. Spread by bark-boring insects, the fungus soon decimated the native elms that had been a primary component of the continent's original forests. Dying robins were merely the latest symptom in this chain of ecological events. It was through a rising awareness of such environmental outcomes that I became distressed by human impacts on nature. This was long before issues of "ecology" were much in the media or public consciousness.

When I graduated at the top of my high school class of 450 students and had scholarship offers at my disposal, I wanted a university where I could study environmental issues with a view toward an outdoor career in natural history or conservation. I was determined to become a professional wildlife biologist.

4 Off to College

The brochure explained that the School of Natural Resources was a relatively small educational unit within the University of Michigan that offered courses on such topics as forestry, conservation, landscape design, and environmental economics. It even tendered a formal degree program in fishery biology and management. Perfect!

Ann Arbor was only about a two-hour drive from Grand Rapids, but otherwise it was worlds apart. The university, with about 25,000 students, opened my eyes to a far grander and more competitive academic, athletic, and social world. The four years spent at this institution were for me a complex time during which dramatic intellectual growth confronted emerging self-doubts, personal and societal issues dueled for attention in my mind, and my growing idealism came face to face with political and environmental realities.

During my freshman and sophomore years, I could hardly wait to complete the mandatory core courses in genetics, physics, calculus, engineering, biochemistry, economics, and the humanities. I had relatively little interest in these topics for their own sake, viewing them instead as necessary hurdles to be cleared before getting to the good stuff on natural history. My favorite classes were small upper-division biology courses with heavy doses of fieldwork, for example, dendrology (the study of woody plants), ichthyology, and fisheries management. Many field trips to local woods or waterways were devoted to identifying the native flora and fauna, and learning about wildlife. One class project, for example, involved generating a topological map of a local pond and censusing its fish to formulate plans for restocking with largemouth bass and bluegill of the proper size and number.

The most enjoyable course of all was aquatic entomology, where I was introduced to the incredible world of larval insects that live beneath the surface of streams and lakes. Unless you've seen them yourself under a dissecting microscope, you can hardly imagine the delicate beauty and variety of these aquatic denizens. Larval mayflies (order Ephemeroptera) display flowery gills on the abdomen and three-pronged tail streamers; stoneflies (Plecoptera) are adorned with two tail filaments and elaborate thoracic shields; larval dragonflies and damselflies (Odonata) come with extendable spoon-shaped mouthparts for capturing prey; various aquatic bugs (Hemiptera) and beetles (Coleoptera) display nifty adaptations hinted at by popular names such as water striders, pond skaters, marsh treaders, back swimmers, water boatmen, water scorpions, riffle-dwellers, and whirligigs; and Dipterans have fine larval forms that belie the less appealing adult mosquitoes and flies into which they metamorphose.

My favorite aquatic insects, however, are caddis flies (Trichoptera). Each larva, a grublike creature about a quarter-inch long, constructs and encases itself in a tubular house of cemented sticks, sand grains, or tiny pebbles. These miniature casings are species-specific in design, and many are as exquisitely beautiful in their own way as the finest of creations by Frank Lloyd Wright. I never cease to be amazed that each *Leptocella albida* grub can cement grains of sand into a finely tapered crescent, that all larvae of *Goera calcarata* somehow know to extend four elongate stones as lateral wings from their central living chamber, or that *Brachycentrus nigrosoma* nymphs consistently build

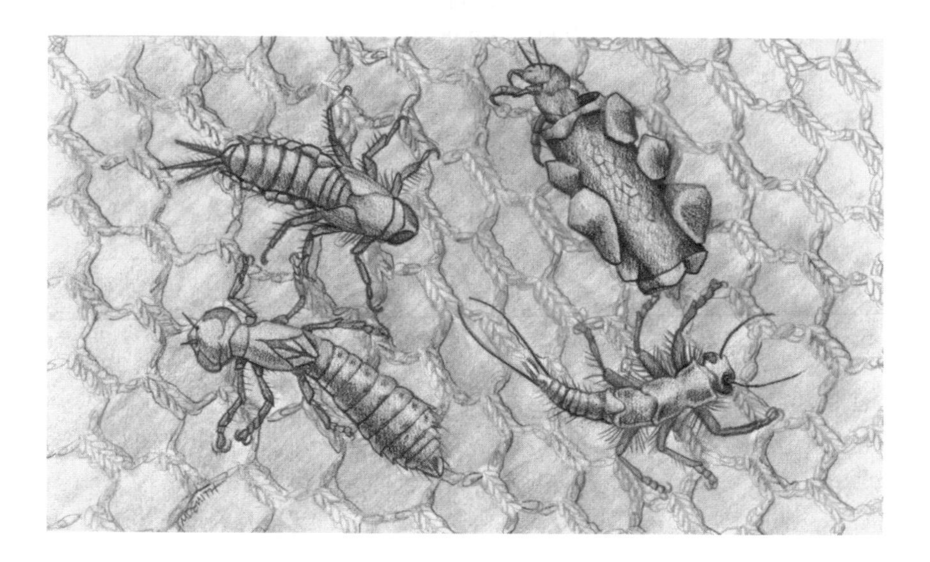

minilog cabins of geometrically perfect pyramidal design. Some species, such as *Limnephilus submonilifer,* incorporate both sticks and stones into the contemporary architectures of their mobile homes. In addition to house construction, other aquatic insects, such as *Arctopsyche irrorata,* have a second building trade: they fashion silky catch-nets to filter food from the stream.

I spent many an afternoon dip-netting and seining in local streams to augment my growing collection of aquatic insects, all painstakingly identified, labeled, and preserved in alcohol. The most memorable of these experiences was a weekend class trip to the Au Sable River, 150 miles north of Ann Arbor. This picturesque river supports an abundance of insect life that contributes in no small part to the stream's well-deserved reputation as one of the nation's top trout streams. Unfortunately, our field trip coincided with a major winter storm. Temperatures plunged to -25 degrees F as we drove the university vans northward through snowdrifts to a log cabin that was to be our weekend home.

Temporarily warmed by the cabin's fireplace, we took turns running down the snowy slope to the river below. There, in cumbersome waders pulled over heavy winter gear, we repeatedly hauled a ten-foot seine through a slushy riffle. At these temperatures, fast-flowing waters tend to freeze from the bottom up rather than the top down. Insect larvae were retrieved from the dripping net even as it froze stiff. Despite the brutish weather, I loved this class outing: a special combination of winter wonderland, camaraderie, and the sheer excitement of the hunt for prized insect specimens.

Michigan is certainly among the finest places on Earth to apprentice in aquatic ecology. The expansive Great Lakes system holds about one-fifth of the world's surface supply of fresh water and is host to a great diversity of underwater life. Lessons on ecological interactions and human environmental impacts abound as well, as illustrated by recent introductions (intentional and unintentional) of exotic species.

For millennia, the Great Lakes have emptied into the North Atlantic Ocean via the St. Lawrence Seaway. In a critical stretch between Lakes Erie and Ontario, the river plunges over Niagara Falls, a formerly impenetrable barrier to natural upstream migration of invasive marine species (and to human commerce on the river). In 1933 a navigable canal bypassing Niagara Falls was completed, and it was to have reverberating biological and economic consequences for the Great Lakes region to this day. Among the first exotic pests to move upstream through the twenty-seven-mile-long Welland Canal (and perhaps through the Erie Canal to the south) were alewife fish and sea lampreys. For a few years, the alewives were held in check in the Great Lakes by lake trout and whitefish. However, people overfished these large predators, and further depredation was caused by the blood-sucking lampreys. Soon, this led to population crashes of the trout and whitefish and to an economic collapse of the multimillion-dollar fishing industry that these and other native species had supported.

In the absence of predation, alewife populations exploded. By the mid-1960s, this species constituted more than 90 percent of the fish biomass of Lake Michigan. Alewives have short life spans and often die in droves, despoiling sandy beaches when their rotting corpses wash ashore. This had serious consequences for the tourist trade, not to mention the beach-front homeowners who had to endure the smell. Furthermore, alewives competed for food and space with the native fish, thereby contributing (along with overfishing and other factors) to the extinction of several deepwater chub species that in prior decades had supported a commercial smoked-fish industry.

Predictably, tourists and commercial fishermen began to raise a stink. In desperation, state fishery agencies conducted a bold experiment by introducing coho salmon from the Pacific Northwest in 1966. Remarkably, the introductions "took." This hatchery program continues today and is viewed as one of the great success stories in the annals of fisheries biology. From an economic perspective, salmon literally brought the Great Lakes back from the dead. Coho and other introduced salmon now support a $3 billion-per-year sport fishery in Michigan that survives, ironically, thanks to the food-chain services of the alewives that previously were an ecological and economic scourge to the area.

Nevertheless, much of the current salmon and trout harvest is indeed for "sport," because high levels of mercury contamination and other toxins caution against more than occasional human consumption of fish from these still-polluted waters. Despite the region's growing environmental awareness and some protective legislation, the Great Lakes continue to be a convenient sewer for much of the region's industrial waste and agricultural runoff.

Nor have the problems caused by exotic species ceased. In 1985 the zebra mussel was introduced inadvertently into Lake St. Clair via the ballast waters of a transoceanic ship. Within four years, this species reached densities of thousands of animals per square meter in some places. Now found in astonishing numbers throughout the Great Lakes, the zebra mussel has had devastating economic effects, clogging the intake pipes of city water supplies, fouling docks and ships' hulls, and in general being an annoyance through its habit of attachment to any hard substrate. Many other exotic species likewise have been introduced via ballast waters or releases of nonindigenous bait by fishermen.

We studied these and many other ecological lessons in our formal course work. The broader messages were clear. Human activities can have profound and often unanticipated impacts (for better or often for worse) on natural environments, even on a grandiose scale. Pollution, overfishing, and introductions of exotic species can result not only in faunal turnover and the extinction of native species, but also can have major economic consequences. The recent history of the Great Lakes showed clearly that decisions based on commercial or engineering considerations alone (such as construction of the Welland Canal) could have unforeseen, long-lasting ecological consequences, often with negative effects on the regional economies those activities were intended to stimulate. In general, it seemed obvious to me that for many reasons, human environmental impacts warranted vastly more attention than they had traditionally received.

These were emerging realizations not only for me personally, but for our society as well. Prior to the environmental movement of the 1960s, there was precious little ecological awareness in the United States (or elsewhere). The word *ecology* itself (coined by Ernst Haeckel in 1866) was employed primarily in scientific circles and would have been foreign to most laypeople. Topics such as clean air and clean water went unraised except by a few courageous pioneers who were quickly denounced as radical hippies or tree-huggers by profiteers of the "military-industrial complex." It fell upon the shoulders of a few environmental leaders such as Paul Ehrlich to alert us to the threats of human overpopulation, Rachel Carson to warn of the dangers of industrial poisons released into the environment, and Aldo Leopold and Garrett Hardin to articulate an environmental ethic. Unfortunately, these were isolated voices, without monetary or political clout.

To appreciate the general lack of social awareness on environmental issues at that time, consider the following evidence from economics. In the 1960s there were no "green advertisements" by oil companies or the timber industry giving even so much as lip service to consumer concerns about clean air, clear water, or endangered species. Such catering was deemed unnecessary (or, more likely, never dreamed of at all) because ecological and environmental issues were not yet a significant part of public consciousness. Today, by contrast, the public's ecological sensibility has increased to the point where many businesses and industries actually see an economic advantage in issuing claims of environmental stewardship (whether bona fide or not) through their public advertisements and reports to stockholders.

In any event, the ecological lessons that I was learning in classes seemed urgent and obvious, perhaps because they dovetailed with intuitive perceptions about nature that I had harbored since childhood. I knew that I had made the right decision in terms of a general career path.

Despite many wonderful times at the University of Michigan, the cloud of the Vietnam War hung ominously over my life. The United States had entered this regional conflict in 1961 and escalated its involvement dramatically during my freshman and sophomore college years (1966–1968). Yet I felt strongly that the war was morally bankrupt. I worried endlessly about how to avoid it, at all cost, when my undergraduate student deferment expired in 1970. I decided that before contributing to what I saw as an unconscionable conflict, I would seek political asylum in Canada or Mexico. Seeking the status of a conscientious objector (CO) was far preferable, but given my lack of formal religious background, it seemed unlikely that it would be granted. Nonetheless, I was determined to explore this proactive possibility as well.

Accordingly, I spent a great deal of my time at Michigan reading Selective Service regulations, researching draft board methods, and assessing legal requirements and application procedures for CO status. I was disappointed to read in the Selective Service Act that the "religious training and belief" prerequisite for CO consideration did "not include essentially political, sociological, or philosophical views, or a merely personal moral code." However, a timely Supreme Court decision in 1970 ruled that a draftee *is* entitled to apply for CO status even if his opposition to war may not be "religious" in the traditional sense of the term. That gave me an opening that was to prove critical.

As a part of the process, each CO applicant is required to meet in person with his local draft board, typically a collection of businessmen and civic leaders unlikely to have much sympathy for "draft dodgers." A formal written statement must accompany the application. The following are verbatim excerpts from a much longer document that I had prepared for my formal appearance before the draft board in the summer of 1970. I wasn't to learn the outcome of this process until many months later.

I have only a brief time on earth. This is the only existence of which I can be sure. Present life may be my only chance for a gift to eternity. For me, it must be treated as the only chance. As such, it is imperative that I live in a manner which I consider loving, good, and helpful to the betterment of mankind and the earth. Whether this can be construed as a religious belief or a moral code of conduct is moot. It holds the same meaning and consequence for me under whatever name. My conscience is the gift of my God through the evolutionary process. To disavow my conscience is to nullify my religion.

As regards the Vietnam war, I am totally opposed to it and my conscience will not allow me to be involved in its continuance. While the avoidance of doing harm is a part of my religion, equally so is my desire to do something positive. The fate of earth is hanging more precariously in the balance than it ever has in its 500 million year history. It will take good will, rational thinking, and love to allow the complexities of life to be perpetuated. I cannot help in perpetuating hatred and violence.

To understand my conscience, it is necessary to appreciate the importance of ecological and evolutionary thinking upon which those beliefs were based. Through the

> study of evolution, I have gained insight into the insignificance of man, which ironically makes what I do in my own life more significant. Through the study of ecology, I have seen the ways in which man must be an integral part of this planet.
>
> These beliefs are not new to me. I determined to pursue studies in natural resources in the hope of strengthening convictions that man and the earth could only survive by harmonious interactions. My training has revealed the turmoil into which the world has been thrown, man against man and against nature. I therefore cannot allow myself to be a part of practices which must, if continued, ultimately destroy the earth.

These were rather heavy thoughts for a naive twenty-year-old, deemed by our society to be eligible to kill other humans but not at that time old enough to vote. They expressed heartfelt sentiments. Indeed, looking back, I am prouder of this episode in my life than in any subsequent achievements. I say this with only the utmost respect for those men and women who did fight in the war and made terrible sacrifices. Most were simply caught up without recourse in a wartime climate of which they wanted no part. Many others were genuinely supportive of the Vietnam effort for reasons perhaps equally as idealistic as my own.

My opposition to the Vietnam War was ideological and firm, but it also came at a personal psychological price that would be difficult to overstate. Throughout my life, I had been a model student and exemplary citizen. But thereafter, my moral sensibilities, rooted in ecological thought, brought me for the first time into direct conflict with the prevailing "patriotic" norms of our culture. Under society's harsh gaze, I felt suddenly transfigured from promising young student to abject coward. Although I felt increasingly alienated and disenfranchised from the prevailing elements of U.S. society, in some respects this only strengthened my resolve.

At first, my stance on the war also created a disturbing rift between me and my father. As a World War II veteran and lifelong federal employee, Dad now seemed from my vantage far too duty-bound to the supposed wisdom of governmental leaders and official national policy. Thus, the whole Vietnam issue caused both of us considerable anguish. I'm not sure whether it was through family allegiance or intellectual reassessment, but to my great relief, Dad finally came to grips with the sincerity of my beliefs and eventually fully endorsed my quest for a CO status.

The draft issue reached a critical stage during my "lottery" year. First came the draft physical, an unforgettable event. In the predawn light of a cold winter's day, several hundred of us college students were bused to Detroit for physical examinations. Stripped nearly naked for the day in a drafty old building (pun intended), we were herded like cattle through examination rooms. Urine samples were collected in unlabeled lots, rectums were hurriedly examined under a doctor's flashlight as we stood bent over in long lines, and psychological profiles were taken by disinterested shrinks. On written IQ tests, which involved multiple-choice questions such as whether a picture of a hammer matched with "nail," "bird," or "foot," many of us filled in random answers. We passed with flying colors.

Another memorable event occurred on the evening of December 1, 1969, when the nation's first draft lottery was broadcast on live TV. It was like an exceedingly sick game show. Draft officials randomly drew 365 numbered balls one by one from a tumbling bin. Each ball specified a different day of the year, thereby determining according to birth date the exact order in which eligible men were subject to military induction. On such luck of the draw hinged the life-and-death fate of hundreds of thousands of young males born between 1944 and 1950.

I remember the evening well. All of us brothers assembled before a television in the fraternity lounge. As each ball was drawn and called out, a sigh of relief erupted from those whose birthdays yielded high numbers, and screaming or sobbing from those less fortunate. We knew that any lottery number above about 200 meant almost certain immunity from military service—an uninterrupted life and career. On the other hand, a number less than about 160 spelled near-certain combat in bloody Vietnam. People with numbers between 160 and 200 might or might not be called up during the year, depending on how much additional cannon fodder would be required to continue a war effort that few people outside of Washington wanted any longer. My number came up 177.

Part Two

Dawn of a Scientific Revolution

Prior to the Darwinian revolution of the late nineteenth century, many naturalists saw as their purpose a glorification of the Creator's works and a revelation of the mind of God by describing and interpreting His biological works through the prism of religion. These passionate natural theologians often provided excellent technical descriptions of nature, but their interpretations of biological phenomena tended to be subjective and doctrinal. As a child, I too had come to revel in the wonder of nature, yet conventional theological motives provided no compass for my natural-history interests. Indeed, I seemed to be in search of some personal, orienting philosophy that might help to rationalize my growing passion for biology.

As I entered the next phase of my life, I gradually came to see biology less as a mere collection of facts and more as a critical scientific endeavor to interpret the workings of nature through unbiased eyes. In particular, I would come to take seriously the "history" portion of natural history. This required adding an appreciation of evolution, genetics, and molecular biology to my earlier training in ecology and environmental studies. Furthermore, given the deteriorating ecological state of the modern world, my hope was that a broader mission might also be served by attempting to integrate understandings from these diverse biological sciences into some sort of conservation ethic. This endeavor would itself become my life's passion.

In graduate school, I fortuitously gained entry into the ground floor of a scientific revolution—the rise of molecular genetics—that would forever change the face of the disciplines of ecology, evolution, and natural history. It was not an easy revolution, however, and even to this day, molecular orientations are belittled by a few ecologists and naturalists who remain unappreciative or even resentful of the perceived hegemony of genetics over traditional organismal biology.

In my own mind, I have come to see no inherent conflict whatsoever between an enchantment with natural biotas, and a fascination with the genes that these organisms inevitably house and transmit to their young. Indeed, my career would become devoted to the notion that the vast historical information content of genes could be tapped through appropriate molecular-level inquiries, and that the genetic sciences could thereby greatly inform our understanding of animal biology and natural history. Only a decade or more after I began to develop this notion, it was to become thoroughly mainstream.

5 Southern Exposure

When I graduated from the University of Michigan in the spring of 1970, my future remained clouded by the Vietnam War. Would my draft number be reached later that year, and would my draft board grant me CO status? Faced with this uncertainty, I enrolled immediately that summer at the University of Texas, assuring myself of onc additional ycar of immunity from the draft via a student deferment. Why Texas? For one thing, I wished for a fresh, anonymous start, free from the sense of social alienation that I had felt in my home state from opposition to the war. Also, I wanted to be in a border state in case rapid escape to asylum in a foreign country became necessary.

Ironically, if not for Vietnam, I would never have entertained thoughts of graduate school. In high school and college, academic life had been of little interest, and I was anxious to start a field career in environmental biology or conservation, perhaps with Michigan's Department of Natural Resources or the U.S. Fish and Wildlife Service. Despite a sterling academic record, I felt neither an attraction nor a genuine aptitude for intellectual competition inside the formal classroom. Learning was enjoyable, but group settings were uncomfortable. Then and now, delivering or hearing lectures in the evaluative context of a formal university course are rather agonizing experiences for me. Any stage presence I may have developed in public speaking is merely a facade. I've always been unsure of my ability to rise upon demand to mental challenges in front of a critical group, so academia was nearly the last occupation I might have envisioned.

Thus, I was caught completely off guard by a gradual transformation to a philosophical appreciation (indeed, love) of intellectual endeavor for its own sake. This adjustment in outlook started with two enthralling graduate courses I enrolled in at Texas: evolutionary ecology, taught by Eric Pianka, and evolutionary genetics, by Robert Selander.

Before then, I had viewed science primarily as the acquisition and application of technical information, be it the cataloging of insects in a stream, managing a pond to maximize fish production, or describing cellular genetic processes such as mitosis and meiosis. Now, in these courses, I was introduced to the notion that the biological sciences also abounded with conceptual issues of broader import. In particular, the logic of evolutionary analysis could be applied to questions of "how" and "why" in addition to "what." Why do small islands show less biotic diversity than comparably sized regions on nearby continents? Why are predator-prey systems often stable or oscillatory? How do new species arise? Why have genetic mechanisms arisen such that males and females tend to come in similar numbers in most species? Why is sexual reproduction so nearly universal? Why are the phenomena of senescence and death nearly ubiquitous to life?

Dr. Pianka, well known for his research on the organization of desert lizard communities and soon to be the author of an important textbook entitled *Evolutionary Ecology,* taught our class how a wedding of evolutionary logic with mathematical, graphical, or statistical modeling could be brought to bear on such questions. For example, we were introduced to the theory of "island biogeography," developed by Robert MacArthur and E. O. Wilson, which interpreted plant and animal diversity on islands as reflecting a balance between rates of species' extinction and immigration. The theory sought to explain, for example, why islands that are large, topologically varied, and adjacent to colonization sources such as continents tend to display greater biodiversity than islands that are small and isolated. MacArthur and Wilson formalized an argument that large islands adjacent to a biotically rich mainland experience high immigration rates and relatively low rates of species extinction, leading at equilibrium to rather large numbers of resident species.

Similarly, we were introduced to Ronald Fisher's mathematical model of sex ratios. Fisher argued that when males are in short supply in a population, families with sons tend to contribute more genes to subsequent generations, whereas when females are in short supply, families with daughters have the reproductive advantage, on average. This form of natural selection operating on individual families tends over time to promote an approximate balance in the total number of males and females in most populations. From this perspective, the evolved genetic mechanisms by which gender is determined (such as the sex-chromosome system in mammals) can be viewed as "proximate" cellular devices consistent with the "ultimate" explanation of equitable gender ratios based on natural-selection theory.

Through detailed appraisals of these and many other biological topics, Pianka convinced our class that evolutionary reasoning was a powerful, indeed indispensable conceptual framework for interpreting nearly all issues in ecology, behavior, and natural history. I had heard the dictum by the famous evolutionary biologist Theodosius Dobzhansky—"Nothing in biology makes sense except in the light of evolution"—and now I was beginning to appreciate why this is true.

Whereas Dr. Pianka's course drew logical connections between evolutionary biology, ecology, and behavior, Dr. Selander's course linked evolutionary science to genetics, biochemistry, and molecular biology. In the mid-1960s, the technique of "protein electrophoresis" had been introduced into population genetics by Richard Lewontin and John Hubby in the United States and Harry Harris in England. In this approach, researchers separate proteins (the functional products of genes) from one another by forcing an electric current through a rubbery supporting medium (gel) that is much like dense Jello in texture. Proteins placed in this gel move through it at varying rates that are a function of each protein's physical characteristics. By revealing otherwise hidden features of particular proteins, the electrophoretic (or "allozyme") procedure enabled scientists to deduce an organism's genotype, its genetic makeup, at each of about thirty to fifty individual genes. The protein-electrophoretic method is molecular child's play by today's standards, but in those days it seemed a mystical biochemical wizardry, and it was to revolutionize the field of evolutionary biology.

The transformation was far from smooth, however. In the 1960s and early 1970s, many ecologists and population biologists were disinterested in or openly hostile to these new molecular approaches encroaching upon their disciplines. Graduate students today may take for granted that population biology and molecular genetics are intimately connected scientific subjects, but less than a half-century ago, there was virtually no contact between these fields. Before 1965, an ecologist or natural historian had no reason to pay special heed to molecular genetics, and a molecular geneticist would not likely have any training or interest in natural history or evolution.

Although I was a natural historian and ecologist at heart, through Dr. Selander's course I began to see the neophyte molecular revolution in a far more positive light. Prior to that era, genetic studies had been confined mostly to a small handful of species, such as fruit flies or Mendel's pea plants, that could be artificially crossed and monitored under controlled laboratory conditions. Now, for the first time, biologists had a procedure that provided ready access to genetic information from any species. I began to see great potential for molecular methods to open the entire biological world to genetic scrutiny. By studying gene products directly in the laboratory, and interpreting them as "molecular markers," much might be learned about natural history, behavior, and evolution.

Dr. Selander was one of the innovators of allozyme analysis in an ecological context, and his course conveyed the intellectual excitement of the times. Many researchers then (and now) became preoccupied with the question of whether the exuberant population-genetic variation revealed in the new molecular assays was promoted by some "balancing" form of natural selection (stemming perhaps from spatial or temporal heterogeneity in the environment, for example), or whether it reflected instead a dynamic equilibrium between the origin of selectively neutral mutations and their loss by random genetic drift. This question interested me, too, and indeed was to provide a motivating rationale for my master's thesis. However, I also began to appreciate a different and less-exploited class of scientific opportunities that allozyme methods afforded.

These (and many other molecular assays developed later) could in principle reveal a plethora of natural "tags." But unlike artificial tags such as leg bands, radio collars, or fin tabs, which researchers routinely employ to monitor the movements of individual birds, mammals, or fish, the genetic tags supplied by Mother Nature are both ubiquitous and heritable across the generations. Thus, properly decoded, various molecular markers might reveal past and present events otherwise unknowable to the field biologist: Who mated with or parented whom? What are the historical patterns of gene movement within a species? What is the branching structure at any depth in the tree of life? Much of my scientific career would be devoted to developing and exploring multifarious applications of molecular markers for ecological and evolutionary issues. By so doing, I was to become a genetic pioneer in a new breed of late-twentieth-century natural historians.

My immediate goals were far more modest. Suspecting that I had only one year at best to capitalize upon opportunities at the University of Texas, I determined to strive for a fast-track master's degree. Toward that end, I loaded my schedule with courses and began thinking immediately about research projects that might lead to a publishable thesis. Dr. Pianka agreed to serve as my academic advisor.

For several weeks, I pestered Eric incessantly about possible research projects. His style, fair enough, was to suggest that I go for contemplative walks in nature with open eyes and mind, and use what I had learned in his course to devise a meaningful set of observations or experiments that would critically address some unanswered ecological question. I took this advice seriously. So, after class each day, I drove to nearby woodlands, fields, and streams, trying to formulate a defensible research project that would mesh natural history observations with my nascent understanding of evolutionary theory.

The city of Austin, home to the university, is adjacent to the Edwards Plateau in an escarpment region that marks a pronounced ecological transition between East and West Texas. To the east are pine-forested coastal plains that resemble portions of southern Alabama or Mississippi as they slope gradually toward the Gulf of Mexico. To the west is the rough karst topography of the hill country, covered by scrubby mesquite, cedars, and prickly pear cacti and traversed by clear streams emerging from limestone springs. As a result, central Texas is at a crossroads between eastern and western (as well as northern and southern) biotas, and many species have ranges that overlap in this area.

Despite my avid exploration of this rich environment, I had trouble devising a thesis project. Many possibilities looked promising at first, but I couldn't translate them into a workable research protocol. For example, I was intrigued by a recent scientific paper showing that sunfishes use species-specific grunting and popping sounds in courtship, in addition to visual cues. I spent considerable time watching bluegills, greens, and other sunfish as the courting males built and defended their nests in the shallow margins of streams near Austin. However, I had no clue as to how to devise critical experiments that might enhance scientific knowledge of these animals' mating behaviors, much less of broader mating system theory.

Frustrated by my lack of research progress, I asked Dr. Selander for ideas. He immediately produced a list of about a dozen significant questions in evolutionary biology ripe for analysis by the newly available allozyme methods. His suggestion brought me face to face with a central enigma of my entire scientific career. I always have considered myself a field ecologist and natural historian, yet I seem unable to devise critical research projects in these disciplines (I greatly admire researchers who are able to do so). Conversely, I never in my wildest dreams envisioned myself as a molecular geneticist, yet here was an emerging field that seemed rife with opportunities to answer fascinating questions about the ecology and natural history of any species.

One question on Dr. Selander's list that piqued my curiosity was whether levels of genetic variation in natural populations were correlated with magnitudes of environmental heterogeneity, as predicted by some population-genetic theories on how natural selection might mold protein variation. Although I didn't relish the thought of laboratory work, here was an executable research project. However, a suitable study organism was needed. Ideally, that species would consist of some populations that had long been confined to stable, uniform habitats, and other populations that inhabited highly variable environmental regimes. Of course, samples from those populations would have to be accessible for the allozyme comparisons.

If proverbial sentiments were to be believed, two habitats on earth are exceptionally stable: the deep sea and caves. Both environments tend to be characterized by uniform darkness, relatively invariant atmospheric or aqueous composition, and constant temperature. Deep-sea organisms were infeasible for me to collect, but caves abounded in central Texas and nearby Mexico. Thus, in the science library, I soon read all I could about biological denizens of the terrestrial underworld. They turn out to be a fabulous lot indeed. Many cave-restricted (troglobytic) species are eyeless and unpigmented, yet come equipped with sensory and physiological adaptations that facilitate survival and reproduction in their pitch-black surroundings. Typically thought to have entered caves as troglophiles (facultative cave-dwellers) thousands of years ago, some troglobytic species are also survived by closely related surface-dwelling (epigean) forms.

A perfect example involved a complex of blind, ghostly pale cave fishes inhabiting underground streams of northeastern Mexico. There, hidden in the limestone caverns of the Sierra de El Abra mountains, these troglobytic fish in the genus *Astyanax* were thought to have diverged in recent evolutionary time from eyed and pigmented relatives whose descendants remain common today in surface streams of the area. Furthermore, in one cave near the village of Valles, adult fish reportedly displayed partial eye development and intermediate skin pigmentation. Was this latter population in evolutionary transition toward troglobytic existence, or might the odd-looking fish be recent hybrids between cave and surface forms? If I could collect the relevant samples, allozyme markers might answer this question and also address the broader issue of whether levels of population-genetic variation and environmental heterogeneity were correlated.

Austin is well known as a haven for spelunkers. The city boasted an active caving club, with many members fanatical in their devotion to cave exploration. I quickly hooked up with one such group, which for years had been mapping newly discovered caves of the Sierra de El Abra. Their protocol (financed on a shoestring by personal savings) was to identify promising limestone sites using small spotter planes and then mount ground expeditions to confirm newly sighted cave openings and map their often extensive interior passages. When spelunkers Bill Russell and Terry Raines learned of my contemplated project on *Astyanax,* they were anxious to collaborate. As luck would have it, they had planned an expedition to the area in about three weeks. That gave them just enough time to get me marginally trained.

The first challenge was to make me comfortable in cave environments. Bill

Spelunkers (bottom center) at the entrance to a Mexican cave.

escorted me into several small caves near Austin, where I learned the importance of backup light sources, how to contort my body to squeeze through passages sometimes little wider than my head, and in general how to mitigate my claustrophobia. These caves were biotically paltry, but I took the opportunity to collect some cute little troglobytic beetles for genetic analysis. The next training lessons centered on rock-climbing techniques that would be needed for the vertical passageways in some Mexican caves. Bill and Terry soon had me rappelling over the edge of hundred-foot-high vertical bluffs above the Colorado River near Austin. The intent was to teach me proper use of ropes and pitons, how to position my body while clinging to a cliff, and how in general to mitigate my acrophobia.

Departure time soon arrived. Loaded down with camping gear, caving equipment, nets, seines, and a dozen large metal canisters for toting fish, we headed south in a dilapidated Volkswagen bus toward the Mexican state of San Luis Potosí. For the next ten days, we roamed the mountains in search of *Astyanax*, exploring caves and surface creeks alike, collecting samples wherever the fish were encountered.

The many caves in that wilderness area defy hyperbole. They are splendid, grand, awesome. If in the United States, they would be a tourist attraction the physical equal of Carlsbad Caverns in New Mexico or Mammoth Cave in Kentucky. Yet the entrances to these Mexican caves lay hidden in forested hillsides, and their whereabouts were known only to a few. Apart from local townspeople and villagers, and an occasional peasant farmer in the hills, we had the region entirely to ourselves.

This whole section of the Sierra Madre Range is honeycombed with underground passages ranging from grandiose to minuscule, such that it is nearly impossible to say where one cave system ends and another begins. Most or all of the network may be connected periodically when the cave system backs up after heavy rain. This possibility was driven home to me at one magnificent cave whose entrance was at the base of a circular sinkhole 200 feet deep and a quarter mile across. From the base of the pit, through my binoculars, I watched a flock of parrots reeling around the rim far above. There, beneath a rocky ledge high on the slope, was a huge log that must have wedged in that position when a deluge had flooded the basin.

Bill and Terry were anxious to explore a remote cave site previously spotted from the air. It wouldn't be easy. The suspected cave entrance was about a two-hour hike from the nearest road, which itself proved to be little more than a rutted foot path through the jungle, barely passable with the four-wheel-drive jeep we rented. After an arduous half-day journey, we actually found the cave.

The effort proved well worthwhile. A vine-cloaked entrance more than fifty feet high opened immediately to an even grander cathedral with huge boulders as pews, recessed ledges as pulpits, and stalactites as organ pipes. The main sanctuary soon branched into corridors. We followed one tunnel for several hundred yards as it alternately constricted and reopened into a series of chambers. I remember my awe at the incredible surroundings and at the thought that we might be the first people ever to have entered this sacred realm. This notion was short lived. At the back of one chamber was a twenty-foot vertical precipice. As we searched along its edge for a way to reach the cave floor below or a site to anchor our rappel ropes, we stumbled upon an old abandoned wooden ladder!

The collecting effort was highly successful. Delightfully exhausted, unshaven, and filthy after a splendid week of camping, caving, and seining, we cruised back to Austin, clearing the border checkpoint at Nuevo Laredo. There, for a couple of hours, guards were certain that they had detained a bunch of drug-running

hippies. We relished the disappointment on their faces when a thorough search of our minibus revealed nothing more interesting than a bunch of weird "minnows" in water jugs. They were soon swimming in neatly labeled tanks back in the laboratory. When Dr. Selander arrived at work the following Monday morning, he was dumfounded to see that his naive young student had pulled off this expeditionary feat. Chest expanded and feigning nonchalance, I secretly felt the same amazement.

With three weeks of furious laboratory effort, I finished the genetic assays. The troglobytic and surface-dwelling forms of the fish proved to be close evolutionary relatives. The genetic markers also revealed that the partially eyed fish were probable hybrids. Furthermore, the cave populations displayed dramatically less genetic variation than their surface-dwelling counterparts, which were among the most genetically variable organisms examined to date. However, several lines of evidence strongly suggested that the diminished variation in the troglobytes was due primarily to the effects of random genetic drift in small, isolated populations, rather than to any selective effects per se stemming from a stable cave environment.

My first plunge into research had been an immediate, unqualified success. This serendipity was pivotally important, more so than I knew. Although a brilliant scientist, Dr. Selander had a short fuse at times and a well-deserved reputation for occasional outbursts, when he would physically boot new students or other visitors from the premises if they didn't produce publishable scientific results soon after their arrival.

While several of us were working in the lab one day, the boss stormed in and ordered everyone out, immediately. So as not to lose the day's efforts, I quickly gathered up my electrophoretic gels and power supplies, put them on a cart, and plugged them back in to electrical outlets in the hallway. Returning to the lab the next morning, we found that during the night a locksmith had changed the locks on the doors and installed a padlock on the refrigerator where chemical supplies were stored. It turned out that Dr. Selander suspected that someone was sabotaging the lab, because many electrophoretic assays had failed that week. I quickly set to work to find the source of the problem. By conducting enzymatic assays and control reactions in test tubes, I determined that the immediate cause was a single bad batch of the chemical NADP, which is used in several enzyme stains. I seriously doubt that any intentional foul play was involved.

Apart from such minor episodes, I never had a serious run-in with Dr. Selander, a rather unusual distinction for students and faculty alike in the Zoology Department. Despite his gruff exterior and temperamental nature, I came to admire and respect Dr. Selander for his powerful intellect and scientific acumen. The only downside to the relationship is that Dr. Pianka, whom I also respected greatly, was upset that I had changed advisors and switched to genetic research.

There are three footnotes to the *Astyanax* study and the paper that resulted from it, which was published in the journal *Evolution*. First, the paper later became a Citation Classic, a rare honor reserved for publications cited frequently in the scientific literature. The study's popularity stemmed from a general fascination with cave organisms and the fact that the project provided one of the first multilocus appraisals of genetic variation in any vertebrate species. This recognition emboldened me to contemplate what formerly had been unimagin-

Descending into a vertical cave in search of blind *Astyanax* fish.

able: that I might be capable of making substantive contributions to the science of evolutionary genetics, a field that seemed impossibly sophisticated to me prior to that time.

Second, although it was only a minor point in the paper, I suggested that the cave and surface populations of *Astyanax* could be considered members of the same species (because they displayed rather high genetic similarity and hybridized when in contact). In the 1930s, Professor Carl Hubbs had placed the troglobytic and epigean forms in separate taxonomic genera, based on their distinctive morphological appearance. In my days at the University of Michigan, where Professor Hubbs had worked years earlier, I had come to idolize this man, who was often called the "Dean of American Ichthyology." Shortly after my *Astyanax* paper appeared in print, I received a personal letter from Professor Hubbs and excitedly opened it, expecting to find a congratulatory note on a fine piece of work. To my horror, it was instead an acerbic attack, concluding that anyone who proposed conspecific status for cave- and surface-dwelling forms of *Astyanax* "must be as blind as the fish themselves." Fully distraught, I destroyed the letter and tried to erase it from my mind, but that phrase was seared into my memory. I wish now that I had saved the letter. I would frame it as a personal reminder of those difficult early years of overt antagonism toward molecular research by some natural historians, whom I otherwise admired greatly.

Third, an offshoot of the *Astyanax* project gave me an early exposure to biochemistry, another field that until then had seemed far beyond my capabilities. During the course of the allozyme assays, I stumbled upon a duplication of an ancestral gene encoding phosphoglucose isomerase. The protein products of these two genes proved to have unusual biochemical features, such as uniting to form functional hybrid molecules within cells and displaying striking tissue-specific distributions in muscle, heart, and liver. With help from a biochemistry professor, Barrie Kitto, I conducted biochemical assays on isolated products of the duplicated genes and showed how they differed in some key kinetic properties. This experience taught me two important lessons: that I could indeed learn sophisticated biochemical technologies and that, nonetheless, biochemistry could not compete with ecology or evolutionary genetics for what I might want in a career.

That year, 1970–1971, Dr. Selander's lab was abuzz with technicians and postdocs doing genetic analyses. One large project involved *Peromyscus* mice, a genus with some sixty species native to North and Central America. This was to be a prototype study exploring how allozyme methods might reveal geographic patterns of population-genetic structure and systematic relationships. By then, several other laboratories around the world were engaged in similar allozyme surveys, but most remained myopically focused on the traditional workhorse of population genetics, *Drosophila* fruit flies, and also on the singular issue of whether the newly uncovered genetic variation was selectively neutral. Researchers had almost completely neglected what Dr. Selander and I saw as another biological frontier. Thus, the door was wide open for a newly anointed "gel jock" such as me to explore an unpopulated research niche, in which genes would be the means rather than the ends of scientific analysis. In other words, I determined to apply genetic markers to questions in ecology, natural history, and evolution traditionally reserved for field biologists and theoreticians. This was the moment when, for me, genetics and natural history merged.

Dr. Michael Smith, a senior faculty member at the Savannah River Ecology Laboratory in Aiken, South Carolina, was visiting Dr. Selander's laboratory that year to facilitate the *Peromyscus* studies. Mike was a leading expert on small mammals, and he was to serve as biological consultant, taxonomic specialist, and collector. A huge man, warm-hearted and down to earth, Mike had a great relish for plunging headlong into the most arduous biological fieldwork. We hit if off immediately.

Mike had been planning an expedition to collect field mice throughout the American desert Southwest and western Mexico. He needed assistance, and two of his graduate students and I quickly volunteered. Off we went in a university van loaded to the ceiling with mouse bait (Quaker oats and peanut butter), rodent cages and water bottles, and 1,000 live traps. Each trap was a spring-loaded box about the size of a small block of cheese, just big enough to accommodate an unsuspecting mouse. Quite unlike house mice introduced from Europe, the native *Peromyscus* species are delicately beautiful rodents with lovely pelage and graceful features of body, face, and tail.

The idea was to capture mice live from numerous sites and periodically ship batches of them back to Selander's lab for genetic analysis. The routine was for each of us to set out 250 traps each evening in long, radiating lines, pick them up the next morning at daybreak, drive about a hundred miles to the next site, and repeat the procedure. On a typical night, about 5 percent of the traps caught mice, about fifty new animals, which had to be marked, caged, watered, and fed as we drove to the next collecting locale.

The trip was a delight. I was exposed to desert environments for the first time and came to love them for their scenic beauty, biotic richness, and sense of solitude, particularly under the clear night sky. The journey took us across western Texas, New Mexico, Arizona, and throughout the Mexican states of Sonora, Durango, and Chihauhua. We collected hundreds of field mice, as well as creamy-colored kangaroo rats *(Dipodomys),* with their long tails and legs built for jumping, nasty-tempered woodrats *(Neotoma),* with their formidable teeth, and various other small desert inhabitants.

One evening, we arrived late at our collection site and had to stumble through the dark to deploy our traps among the boulders and prickly scrub of what seemed to be a suitable hillside. In morning's light, we discovered that we had laid our traplines among stones that spelled out a large political advertisement visible from the town below. In front of curious villagers, we sheepishly picked up our traps and departed. On another occasion, camped in a wilderness area, I was awakened in the predawn to the sound of metallic clanking. It turned out that a fourteen-year-old shepherd had stumbled upon our trap lines and was depositing the strange metal boxes into burlap sacks on his burro's back. I doubt that the boy understood our feeble explanation of why the traps were so important to us.

One night, after we set traps and had a celebratory beer or two, I accidentally slammed the van door on my finger. I couldn't sleep that night, and in my dazed state committed the cardinal sin of using tap water to swallow aspirin. Diarrhea struck the next morning. Forced to hold my throbbing hand over my head to alleviate the pain, and wracked by stomach cramps, I must have been a sorry sight indeed as I struggled through the desert landscape to retrieve traplines whose whereabouts only I knew.

My year in Texas was a time of considerable personal growth. I was exposed to new lands and biotas, and more importantly, to a fundamentally different perception of the broader art of scientific inquiry. Through well-mentored exposure to the conceptual side of evolutionary biology, and to one technical side of molecular genetics, I had come to love scholarly issues that in prior years I might have viewed with trepidation or naive disdain. For the first time, my eyes were opened to a wealth of engrossing intellectual challenges, and better yet, these seemed to mesh well with my inherent interests in ecology and natural history. Although I could never have imagined it in earlier years, I was now extremely anxious to continue graduate studies. Unfortunately, that would have to wait: about halfway through the academic year, my lottery number in the draft came up.

6 Heading East

Initially, the news was devastating. However, far better tidings also arrived in the mail. I had been granted conscientious objector status! I held no illusions that this reflected any exceptional religious or moral insights on my part. More likely, the draft board's decision was in recognition of my academic achievements in an ecology curriculum deemed consistent with my personal ethical stance. Being awarded the CO spared me from the conflict in Vietnam, but it also deepened my resolve to honor the social commitment upon which the award was based: to strive during my life to make genuinely positive contributions to environmental issues.

The immediate task was to find suitable alternative service. Under Selective Service guidelines, a CO whose number comes up is granted a short time to identify a two-year job deemed by the draft board to be conducive to the "national health, safety, or interest." The salary must be at or below that of a military inductee, and it helps if the position entails physical danger or discomfort. If the CO is unable within the allotted time to find such a job, the board assigns one—typically, as a janitor in a Veterans Administration hospital or an ambulance driver.

In my case, alternative service had to commence before summer's end in 1971. At Texas, I hurriedly made inquiries and tentative plans for possible positions. Most of these fell through. For example, hoping that I might be able to serve in the Peace Corps, I used meager savings to fly to Chicago for an orientation session on overseas openings, only to learn that government regulations deemed the Peace Corps unsuitable for alternative service.

Then, an opportunity appeared. During his sabbatical in Texas, Mike Smith had been impressed with my skills in the lab and field, and he asked whether I would work as his technician at the Savannah River Ecology Laboratory (SREL) near Aiken, South Carolina. This ecology facility is situated on the Savannah River Plant (SRP), a 300-square-mile federal reserve then operated by the U.S. Atomic Energy Commission. Fenced, guarded, and closed to the public, the SRP contained several nuclear reactors dedicated to producing radioactive tritium, used to make the nation's atomic weapons.

The primary mission of SREL was to monitor the biological effects of thermal effluents and radioisotopes released from the nuclear reactors into the local environment. Toward that end, biological samples of all sorts had to be collected and tested at regular intervals from various SRP sites, including the infamous swamp. This snake- and alligator-infested backwater of the Savannah River receives several creeks, whose waters were diverted to cool the radioactive cores of the nuclear reactors. Mike envisioned many relevant field projects and collecting tasks for me. He also wanted to capitalize upon my background to set up a protein-electrophoretic facility so that SREL scientists could monitor the genetics of local animal populations.

Mike wrote a wonderful letter to my draft board emphasizing how this position related to national security and interest. He also assured them that my pay would be minimal and would include no room or board, and that my hire would save taxpayers considerable money compared to contracting someone with comparable skills. Furthermore, he made it clear that the position involved physical discomfort, if not risk. The members of my draft board accepted the arguments and approved the position. I've often wondered if in the back of their minds they might have been thinking, "Hah, we'll get that troublemaker—let's throw him to the alligators." In any event, I was as delighted as Brer Rabbit being thrown into the briar patch.

The SREL position was to start in late summer. Meantime, Dr. Selander asked me to accompany him to the Marine Biology Laboratory (MBL) at Woods Hole, Massachusetts. Each summer, MBL invites prominent visiting scientists to conduct research at the facility, interact with the resident staff, and teach graduate students. As an invitee that year, Dr. Selander enjoined me and his postdoc, Walter Johnson, to escort him to Woods Hole to set up a temporary allozyme lab. I could also take one or two of MBL's famous summer courses. The offer was too good to pass up.

At Woods Hole, I took courses in microbial ecology and general ecology. The latter was cotaught by E. O. Wilson, of *Island Biogeography* fame, and who later gained prominence by writing acclaimed books on sociobiology, biodiversity, and human nature. I love these eloquent books and how they demonstrate that important scientific issues can be brought to the public. Wilson always manages to interweave a deep appreciation of intellectual endeavor with a heartfelt concern for the natural world, and he has long been a model for me.

On the research front, Walter and I scurried to set up a working lab from the boxes of gel rigs, power supplies, chemicals, and other materials that we had shipped from Texas. We also chose a research topic. Shortly after our arrival, we noticed that fiddler crabs were prominent inhabitants of the local waterfront. Behaviorists working on fiddler crabs elsewhere had observed variations in the ways that males wave their one oversized claw during courtship displays. Some of the behavioral differences were suspected to register "sibling species," which look alike but are reproductively isolated. We reasoned that allozyme methods would permit us to uncover genetic patterns of geographic variation in fiddler crabs and perhaps identify new species as well.

At Woods Hole, two described species of fiddler crabs could be observed scrambling about in large herds over the marshlands: *Uca pugnax,* which prefers muddy substrates for its burrows, and *U. pugilator,* which favors sandier turf. During that summer of 1971, we collected and genetically analyzed a total of 2,800 specimens from thirty-five locales along the Atlantic seaboard. Most came from nearby sites on Cape Cod, but distant collections were donated by other researchers.

The research collecting as well as class activities at MBL enabled me to explore the beautiful Cape Cod environs. The town of Woods Hole is situated on a southern tip of the peninsula, just a short ferry ride from the islands of Martha's Vineyard and Nantucket. About fifty miles away is Provincetown, at the tip of a curved sand spit that gives the Cape its characteristic hooked shape on a regional map. On the inner side is Cape Cod Bay, bordered on the south by Barnstable Marsh, which was a favorite sloshing ground for ecology field trips. Here, I was taught the ecological significance of marshlands, not only because of their understated beauty, but also as critical nurseries for the larvae of many ocean fishes, shellfishes, and other marine creatures.

The MBL is famous for its long-standing emphasis on neurobiological research. Seemingly an odd topic for a marine lab, this makes more sense when it is appreciated that nerve axons of squid are a model system for neurological experimentation. Axons are the physical extensions of a nerve cell that carry impulses, and squid have unusually large ones, which can be easily studied. To support the demand for fresh squid, one of MBL's research vessels returned to the dock twice weekly with specimens. The boat also brought in an assortment of fish, including herring, midshipmen, tomcod, eelpouts, pipefish, seatrout, croaker, cutlassfish, mackerel, sea robins, tautogs, puffers, flounders, and tonguefish. These were made available to any MBL researcher who could put them to use.

As part of the gene-duplication project I had initiated in Texas (on the metabolically important protein known as phosphoglucose isomerase), I now wanted to trace the evolutionary origin and history of the two functional daughter genes I had discovered. So, I assayed each new fish species that arrived at the MBL dock for presence versus absence of the duplicate genes and their patterns of expression in different tissues. The two genes proved to characterize most fishes, but only one copy was present in most other vertebrate groups. Through such evolutionary-genetic detective work, I discovered that the ancestral gene must have duplicated more than 150 million years ago, followed soon thereafter by functional specialization in different fish tissues. This research led to my second scientific publication and, more importantly, to a growing confidence in what I was doing.

This splendid summer in New England was closed by two memorable events. A happy experience was a cookout by the ecology class, the meal consisting of locally caught seafood items that we had studied in the course. Each team of students prepared a different delicacy. My group chose silversides, tiny fish caught by the hundreds in seine hauls along the beach. When battered, deep-fried, and munched whole, they were a crunchy delight. Even better were butter-sautéed periwinkles (sea snails) attractively served over a decorative bed of seaweed. To fill out the menu, we also savored pit-cooked oysters, lobsters, and other traditional clambake fare.

A less enjoyable event was unexpectedly thrust upon me by Dr. Selander. Forced to leave Woods Hole early to tend other business, he left me to present our group's research findings to a summers-end symposium of students, researchers, and staff. This was my first scientific talk before a large audience, and I was very nervous. Fortunately, I had exciting findings to report. We had discovered a new species of fiddler crab, and through detailed statistical analyses of geographic variation had adduced strong evidence for the influence of natural selection on some of the allozyme markers. My presentation went well, but I was uncomfortable during its delivery and drained afterward (feelings that accompany all of my lectures to this day).

Within a few days, I was on my way to the Savannah River Ecology Laboratory and my first significant exposure to the southeastern United States. I rented a tiny two-bedroom house in Aiken, sharing it with another starting employee at SREL. Each day, Dick Geiger and I commuted to the laboratory, a twenty-minute countryside drive in my faithful but dilapidated 1966 Ford Mustang convertible.

In many respects, Aiken was not a typical small southern town. Many of its citizens worked for the Atomic Energy Commission at the Savannah River Plant and had engineering or Ph.D. degrees. Also, the town was a popular winter residence for a gentry of racehorse enthusiasts from New York and other northern states. Although I was dirt poor, I could vicariously enjoy the rich ambiance of the Aiken surroundings: beautifully landscaped mansions, pastures, and stables; weekend foxhunts, in which splendidly regaled horses and riders chased the hounds through a township preserve; and the annual race-day festivities, when the whole town turned out to watch equestrian contests.

Notwithstanding its imposing nuclear reactors and cooling towers, the SRP site itself is a natural expanse of pine and deciduous forests, pastures, and swamps. Traversed by clear streams, it is also dotted by natural wetlands, known as Carolina Bays, and by several reservoirs, including 1,100-hectare Par Pond. I became enamored of the region and its wildlife. Even today, when I revisit the SRP and drive its uncluttered roads, I am amazed at the abundance and diversity of animals: deer, bobcat, feral hogs, birds of many species, lizards, snakes, frogs (especially evident on rainy nights), and turtles. The southeastern United States is a hotspot for fishes and aquatic herpetofauna on the continent, and SRP must be its core. The diversity of SRP's invertebrate life is also incredible. Three Runs Creek, for example, reportedly contains more species of aquatic invertebrates than any comparably sized stream in North America.

During my two years at SREL as a research technician, I took full scientific advantage of the site's rich animal life. There were seven senior-level ecologists on staff and about a dozen postdoctoral associates and laboratory technicians in the various research programs. Nearly every week brought some new adventure as one or another of these groups called upon my assistance in their ongoing fieldwork.

One long-term project involved surveying the genetics and ecology of white-tailed deer, a species of special concern at SRP because of collisions with cars. Accordingly, SRP officials conducted regular hunts to cull the herd. For each one-day hunt, about thirty sportsmen selected by lottery from surrounding counties were escorted onto the SRP site. In a choreographed ritual, they were driven to a designated acreage and stationed at 200-meter intervals around its perimeter. Teams of hunting dogs were released, and the flushed deer were shotgunned as they fled the woods. The Forest Service patrolled the perimeter by pickup truck, gathering the carcasses and moving them to a central processing area staffed by several of us SREL workers. Our job was to gut each deer and preserve tissue samples for later genetic analysis. At day's end, the dressed-out deer were released to the hunters.

A typical hunt brought in about thirty to eighty specimens, so those of us in the processing area were soon up to our ankles in deer blood and guts. It was an appalling scene. We hoisted each retrieved carcass onto a rack, gutted it, and froze heart, liver, muscle, and blood samples in liquid nitrogen. Entrails and the smelly contents of rumens accumulated in a large mound, and some of the pack dogs abandoned the chase in favor of bared-fang fighting over the pile. Still, I had to remind myself, this gory scene was nothing compared to what was taking place daily in Vietnam.

Many collecting activities took place in the biotically rich swamp. One senior ecologist on SREL's staff, Dr. Whit Gibbons, is among the world's leading experts on reptiles and amphibians, and an indefatigable natural historian. Fearless in the face of alligators or rattlesnakes, he tried to instill in me a similar insouciance. Whether we were up to our waists in swamp mud at midnight grabbing *Hyla* tree frogs in the glow of headlamps, cruising roads during a thunderstorm in search of "herps," or rummaging through abandoned piles of corrugated sheet metal that might be home to a special snake, Whit was always ebullient and, well, reckless.

My fondest memory is of a nighttime collecting trip, canoeing down the swift waters of Three Runs Creek, deep into the swamp. Our quarry were cottonmouths and water snakes. The former are poisonous, but the latter actually have a nastier disposition and can deliver painful bites that cause infection. With me in the bow, Whit in the stern, and headlamps lit, we careened down the river, trying to spot snakes in the overhanging bushes. They were quite common. Pulling up to an occupied bush, Whit would grab a snake with bare hands and thrust it into a burlap bag. This posed little difficulty except one time, when two large specimens were discovered in close proximity. Whit threw one snake into the bottom of the canoe and then grabbed and stuffed the second into a bag. I was just a tad anxious as I waited, feet raised high, for Whit to retrieve and sack our other slithery shipmate.

Much of the fieldwork at SREL involved tag-and-release projects. One ongoing wintertime activity was to band American coots (ducklike birds) on the open water of Par Pond. The capture procedure was eccentric but effective. Using SREL's twin-engined Boston whaler, we would speed full-throttle toward a flock, then focus like a prudent predator on a particular individual. A coot's feet are far back on its body, so it must paddle furiously across the water before gaining takeoff speed. With the boat rapidly bearing down, a taxiing coot usually aborted the takeoff and dove underwater instead. A few seconds later, he popped up nearby and tried to fly but again was thwarted as we bore down hard. After two or three rounds of this exercise, an exhausted coot was netted, banded, and released. Later observations of the leg bands provided information on the birds' movements and migrational patterns.

One such chase brought us more exhilaration than we expected. With Mike Smith at the helm and me in the stern, we sped toward a coot who decided to flee shoreward. Focused intently on the bird rather than where we were going, Mike noticed our danger only at the last moment and cut the boat hard to port. As the hull dug in sideways, Mike was catapulted from the boat in a beautiful forty-foot, 360-degree cartwheel, splashing down just a few feet from shore. Thrown into the gunwale and dazed, I nonetheless managed to grab the steering wheel and slack the throttle. Regaining my senses, I circled back to retrieve Mike and found him shivering and sheepish, but miraculously unhurt. I rated his dive a perfect ten!

Another tagging project, headed by Tom Murphy, involved placing radio collars on Par Pond's alligators. The adorable young were easy to catch at night. From a boat, we just spotlighted the telltale green eyes along the shoreline and jumped out to grab the foot-long creatures. Their parents provided a greater challenge. The three-person procedure, which I learned under Tom's tutelage one remarkable night, required only a small rowboat with an electric motor, a long rope coiled on the seat, a towel, duct tape, a cane pole, flashlights, and a dash of human bravado.

On a black, dead-calm night, we quietly cruised the pond, searching for green reflections in the spotlight's halo. We quietly followed a pair of florescent eyes swimming about ten meters ahead of our boat on the water's surface. The eerie green dots finally permitted us to pull alongside, whereupon we used the cane pole to slip a noose over the creature's snout. Then, all hell broke loose. The seven-foot gator spun and thrashed, wrenched the pole from our hands, and splintered it to shreds. The coiled rope peeled out to its full length, and the behemoth towed the boat around the pond like a small owner on a Doberman's leash. Finally, the alligator tired and dove to the bottom. This was our cue to row into shore, attach the rope to a sturdy tree, and winch in the protesting animal inch by inch.

When the gator's head finally emerged onto the shore, we gingerly tossed a wet towel over its eyes to elicit a quiescent state, which mostly reflected the animal's exhaustion. Then, as a team, we jumped it—one person straddling the base of the tail to keep it still, me sitting astride the animal's back and wrestling its forelegs into a hammerlock, and Tom sitting on its head, securing the jaws shut with rope and tape. After the animal was tied and immobilized, we hoisted it into the boat for the return trip to the dock. The next day, we released the gator unharmed with a stylish new radio collar that would send out information on its comings and goings over the next several months.

Par Pond is renowned for its large fish as well, including abundant large-mouth bass, bluegill, and crappie. Their fast growth rate in waters heated by the nuclear reactors, combined with the lack of fishing pressure, resulted in a bounty of fish the likes of which you've never seen. With fishing rods and tackle boxes in hand, we often headed out in the evening to catch that month's required sample of fifty specimens for genetic and radioisotopic analysis. Almost every cast

Seining bluegill sunfish from a Georgia stream (the author is on the right).

brought in a near-trophy specimen. Indeed, the only sporting challenge was to catch two fish at a time. If a large lure was used, the part left dangling from the mouth of a hooked fish often was attacked by a second.

Much of my work in the new genetics laboratory involved protein-electrophoretic studies of these animals. One mission-oriented issue was whether the heated waters and/or radioactivity had altered the genetic composition of fish populations on the SRP site, a question that also required "control" collections from other geographic locales. One large survey involved sampling bluegill sunfish from nearly 100 localities from the Carolinas to Texas. This genetic study revealed two highly distinct forms of bluegill in the region, one native to the Florida Peninsula, and the other to Gulf Coast drainages. These two types proved to meet and hybridize in a broad contact zone across Georgia and South Carolina. We also found that population samples within each lake usually were genetically similar, those from separate reservoirs differed significantly in genetic composition, and fish from the SRP showed no overt genetic signs of having been grossly altered by exposure to the effluent of nuclear reactors.

Through our extensive collecting efforts, I came to know the rural South well, and bluegill habits intimately. For example, one project involved sampling forty bluegill from each of sixty-four localities: eight sites evenly spaced around each of four reservoirs within each of two parallel drainage basins. We seined repeatedly at each site until the sample was obtained. In a few difficult cases, we resorted to alternative collecting methods: electroshocking, using a backpack rig or a special boat with generator and electrodes; or chemically treating the water with rotenone, which briefly deprives fish of oxygen and causes them to float to the surface.

In the summers, these field expeditions were welcome breaks from laboratory work, if for no other reason than wading neck deep in rivers and impoundments was a great way to keep cool. In winters, we donned wetsuits for warmth. Once, while collecting in Mississippi, a cold front moved through, and we had to break through a veneer of ice on one lake before pulling the seine. On such trips, it was difficult in a cramped van to remove our wetsuits after each seining stop, so sometimes we didn't bother. Once, when dressed only in such rubbery attire, Mike and I strode into a McDonald's restaurant for lunch. Mike told me not to feel embarrassed—although our black suits looked a bit odd, our money was just as green as anyone else's.

In my allozyme studies at SREL, I came to appreciate that historical as well as spatial perspectives are critically important for proper evolutionary-genetic interpretations. This realization led to my first attempts to decipher the historical relationships among allied species by examining protein data. For example, using allozyme methods I compared the proteins of several sunfish species to learn more about how these fishes are related to one another, and how many genetic changes might have accompanied the process of speciation.

My old friend the bluegill is one of about a dozen species of *Lepomis* sunfish, several of which often produce viable hybrids in nature. Did these species differ greatly from one another genetically, or was the fact that they retained the ability to hybridize indicative of an unusually close genetic relationship? The empirical answer proved clear: in protein-electrophoretic assays, each pair of sunfish species differed genetically at a large fraction of their genes—far more so than humans and chimps, for example. Mike Smith and I analyzed and began to published these findings in papers that were among the earliest of their ilk in an allozyme era that was to reach its apogee in the late 1970s and early 1980s.

Another rich vein for scientific mining took Mike and me back to Texas for a short visit. Dr. Selander had the bad habit of neglecting older data sets whenever he became preoccupied with a new pet project. During our year in Austin, Selander's lab had accumulated a shelf full of loose-leaf notebooks housing photographs of electrophoretic gels from mice that we had collected in Mexico. For two weeks, Mike and I labored through the notebooks and scored thousands of genotypes from the gel pictures. Back at SREL, I analyzed these data sets and wrote them up into a series of papers that were a great boon to my developing career.

One of these genetic studies analyzed patterns of geographic variation in deer mice across North America. Another used genetic information to deduce the historical origins of desert mice on various islands in the Gulf of California. For example, based on their genetic relationships, three mice species endemic to the islands of Salsipuedes, Tortuga, and Angel de la Guardia each must have originated fairly recently following colonization of the islands by another species of mouse widespread in Baja California (rather than mainland Mexico). The idea was not lost upon me that these kinds of genetic analyses went well beyond MacArthur and Wilson's equilibrium approach to island biogeography in the sense of revealing not only general diversity patterns, but also the idiosyncrasies and historical specifics of each island colonization event. In general, these genetic analyses of wild populations were among the first of their type in those early frontier days of molecular-genetic exploration of the natural world.

My two years at SREL were a delight. Yet, as time wore on, yearnings for further intellectual challenge made me anxious to complete the CO requirements and return to graduate school for a Ph.D. degree. I had become enthralled by my early experience combining genetics and natural history, and was hungry to learn more. With strong academic and research credentials documented in publications from Texas and SREL, I felt confident that I would be welcomed by many university graduate programs.

7 California Dreaming

In returning to graduate school, I wanted to live in the only part of the country I had not experienced—the West Coast. I soon narrowed my choices to the University of California at Davis and U.C. Berkeley. These campuses had prominent programs in genetics and molecular evolution, and they featured on their respective faculties Francisco Ayala and Allan Wilson, two esteemed leaders in these fields. I was accepted by both institutions, so I studied information on their graduate programs and campuses, trying to decide which to attend.

Berkeley, I read, was situated near San Francisco and Oakland. This was a negative, because I've never cared much for big cities. Davis, the brochure noted, was a small town nestled in the Sacramento Valley between the Coast Ranges to the west and the Sierra Nevadas to the east. I imagined the campus in a picturesque alpine valley with mountains towering on either side. Accordingly, I chose Davis. Upon arrival, however, I realized that the agricultural Sacramento Valley is a good bit flatter and wider than I had supposed. Only on the clearest days can mountain peaks be seen as tiny blips on the horizon. Berkeley, on the other hand, overlooking scenic San Francisco Bay, is one of the most beautiful campuses in the world. I learned quite a bit about brochures that year.

In Francisco Ayala's laboratory, I joined a relaxed yet intellectually stimulating group that included several other graduate students, postdocs, and research technicians. Another important member of the group was Theodosius Dobzhansky, the famous geneticist who had been responsible for weaving the empirical and experimental sides of population genetics into "the modern syn-

Members of the Ayala laboratory, 1975. John Avise is in the center of the top row, flanked to his right by Theodosius Dobzhansky, to his left by Francisco Ayala (in glasses), and behind Olga Pavlovsky. Other people mentioned in the text are Lori Barr (far right) and John McDonald (bottom row, second from the left). Photo courtesy of F. Ayala.

thetic theory of evolution," an intellectual synthesis that stands as one of the great scientific accomplishments of the twentieth century. Dobzhansky, affectionately known to us as Doby, was now of emeritus status. Following his wife's death in 1968 and his retirement in 1970, he had accepted an invitation from Francisco to accompany him in a move from Rockefeller University. Still mentally alert and scientifically astute, Doby maintained a small research lab adjacent to Francisco's, and he was an inspiration to us all.

Dr. Ayala's lab, like those in Texas and South Carolina to which I was accustomed, was geared to run large numbers of protein-electrophoretic gels. Francisco's program centered on the population genetics and evolution of fruit flies. One ongoing project involved monitoring changes in gene frequency in a local population of *Drosophila pseudoobscura,* and Francisco's students had the monthly responsibility to collect specimens from a university-owned ranch in the Coast Range, about an hour's drive from Davis. I enjoyed these opportunities for an occasional afternoon break amid lovely hillsides blanketed by grasses and scattered oaks.

A more extended outing took place when Francisco asked me to accompany him on a two-week expedition to eastern Mexico. We flew to Mérida, where we rented a jeep that took us through the Yucatán Peninsula and into Belize via the only road entering that small country from the north. Collecting fruit flies bears little resemblance to tagging alligators or seining bluegills. Indeed, it's positively civilized. You simply buy ripe bananas or plantains in a local farmer's market, squash them with baker's yeast into a thick soup, place plastic buckets of the broth in strategic field locations, and go sip a martini. An hour or two later, you return with a net in hand, swish it over each bucket, and bottle the captured fruit flies. A nice perk of such relaxed collecting is that it leaves plenty of time for sightseeing. We toured fabulous Mayan ruins and wandered the nature-rich, vine-tangled rain forests of lowland Belize.

Another perk of working in Dr. Ayala's lab was that each graduate student could select his or her own thesis topic on any organism. One of my fellow students, Dennis Hedgecock, used genetic markers to study migrational behaviors of *Taricha* salamanders, bright orange- and red-bellied amphibians that were conspicuous inhabitants of local ponds and streams. Another student, John McDonald, worked for a time on barnacles, which are abundant on rocky Pacific shores.

This reminds me of an amusing incident that really wasn't so funny at the time. A severe hurdle in conducting research on natural animal populations is that scientific collecting permits frequently are required by federal, state, and local regulatory agencies. The law-enforcement arms of these bureaus often have little understanding of science and even less sympathy for academic biologists. Although it seemed unlikely that government officials would be concerned about the death of a few barnacles for genetic research, John decided to play it safe and drive the forty minutes to the state capital to fill out a collecting application. On his way out the door, he asked me to cover for Dr. Ayala (who was overseas on a business trip) should anyone from Sacramento call to confirm John's student status. In my usual absentminded way, I quickly forgot John's request.

Meanwhile, in Sacramento, John dutifully filled out a permit form that required a professor's certification. John left for lunch, forged Dr. Ayala's signature, and returned to the permit office an hour later. Sure enough, a suspicious bureaucrat behind the desk called our lab and asked to speak to Professor Ayala. Answering the phone, but busy and distracted, I informed the unidentified caller that Professor Ayala was out of the country for the month.

The next thing poor John knew, he was formally charged with a misdemeanor—falsifying state documents—and a courtroom date was set. I was not among those John invited to serve as his character witness! Other friends later recalled that fateful morning in court. Seated among prostitutes, pimps, and disheveled burglary suspects, John sat dressed in a neatly pressed, borrowed suit. His was about the tenth case on the docket. The bureaucrat from the permit office made his arguments, and John issued an apologetic defense, saying he never imagined that sacrificing a few barnacles in the cause of science could have created such trouble. The judge listened incredulously, heard the character witnesses, and then issued a decision—the *plaintiff* (not John) was to be reprimanded severely for wasting the court's time!

This incident notwithstanding, we members of the Ayala lab soon bonded into a tight-knit group. The camaraderie was strengthened by a good-hearted feud between our genetics group and several researchers in a biochemistry laboratory housed in the same building. Biochemists, it seems, love to impugn population geneticists for practicing "soft" science. They jokingly chided, for example, that we used only crude extracts from liver or muscle tissues in our allozyme assays. "How can you learn anything," they asked, "if you don't first purify your protein?" That comment bothered me to some extent, until one day I thought of an effective retort. The first information we uncover in our electrophoretic assays is an individual's genotype. "How," I asked, "can you biochemists consider your methods so superior when you don't even know the homozygous versus heterozygous state of an animal from which you purify a protein?" That comment silenced our smarty-pants friends, but only briefly.

Actually, they did have a valid point about our lack of sophistication on matters molecular. For several months, John McDonald, I, and Moritz Benado (another graduate student) conducted a side project that required use of a spectrophotometer and other specialized equipment. Trotting every day to the biochemistry facility, donned in our white lab coats, we were viewed by the locals as the visiting Three Stooges, an impression dangerously close to the mark, as evidenced by an embarrassing incident. One day, I was trying unsuccessfully to weigh chemicals on an unfamiliar balance. Seeing my difficulties, a resident graduate student asked, "Did you tare the weighing paper?" Having never heard the term "tare" before, and thinking he meant "tear," I replied in all seriousness, "No, I cut it with a pair of scissors." The biochemist nearly bent double with derisive laughter.

Still, our experiences in the biochemistry laboratory weren't all negative, and I finally got sweet revenge. Several years later, I married away their beautiful and intelligent technician, Joan. However, Joan can't entirely escape her biochemistry background: she still thinks of me fundamentally as a buffoon.

Davis is a world-class bicycle town, and each day I pedaled the flat three miles between school and a small apartment that I shared with Bill Emery, another graduate student, who was only slightly better off financially than I. My sole income was a National Science Foundation scholarship that paid $200 per month, barely enough to cover the $120 rent and a self-budgeted allotment of $15 per week for food. Joan, being gainfully employed, was wealthy by comparison, so she typically paid her own way on our dates and supplied transportation in her Chevy Nova.

Bill and I ate a lot of cereal, but we also discovered a trick that kept us in protein despite the tight budget. A few miles west of town, Putah Creek emerges from the foothills of the Coast Ranges to traverse the Davis campus before flowing into the Sacramento River. The foothill portion of the creek was alive with stocked rainbow and brown trout as well as native crayfish. After work, Bill and I often drove to fishing sites on the river or to Lake Berryessa, further upstream, where we usually caught our limit of trout and filled a bucket with crayfish.

At school, I enrolled in several courses in genetics and evolutionary biology. One that had special influence on my scientific development was taught by Dr. Les Gottlieb, well known for his research on the biochemical genetics of plants. There, I was introduced to an uncompromising approach, wherein all scientific data and conclusions are viewed with utmost skepticism. Our class read, dissected, and then lambasted the experimental designs or interpretations in paper after paper from the scientific literature. Even the best of research has methodological or conceptual shortcomings when scrutinized closely. With practice, it's easy and even seductive to get into this hypercritical evaluative mode. We learned that such an analytical posture is necessary for truly penetrating analysis, and ultimately for scientific advance. On the other hand, I also came to appreciate that jaundiced attitudes toward exploratory research must be tempered, lest nothing be ventured.

On the dissertation front, I had several false starts. In one contemplated project, I would examine the genetics of speciation in a spectacular evolutionary "flock" of about twenty closely related minnow species endemic to Lake Lanao, in the Philippine Islands. A long-standing evolutionary puzzle was how and when such a great diversity of species might have arisen within a single body of water. However, a friend doing water-quality research on the lake sent word that most of the native fish in this extraordinary evolutionary theater were now extinct or extremely rare due to overfishing, pollution, and human introductions of exotic predators. Such environmental travesties, all too common in the modern world, are depressingly tragic to us natural historians, yet often go completely unnoticed by broader society.

Another project would have compared the genetic structure of freshwater versus marine populations of threespined sticklebacks. I made preliminary collections of these common West Coast fish, but the protein gels ran poorly, and I finally abandoned the effort. As an outgrowth of this work, I crossed sticklebacks in the lab by stripping eggs and sperm from ripe individuals and raising the resulting broods in fifty small aquaria. I wanted to examine the genetic basis of different forms of body armors that protect the sticklebacks against predators. This work led to a brief paper, but the main lesson I gained from the whole exercise was how difficult it is to obtain genetic data via traditional animal crosses. Indeed, a productive protein-electrophoretic laboratory might obtain in a single month far more genotypic information on virtually any species in nature than had accumulated from conventional breeding experiments in all the 100 years since Gregor Mendel.

The dissertation project that finally succeeded began with allozyme analyses of about ten species of minnows (family Cyprinidae) native to California. Initially, I encountered some of these species on collecting trips for sticklebacks, and they proved to be an interesting lot. Unlike the small forms back East, several West Coast minnows are top-level predators. The squawfish can grow to more than a meter in length, and large specimens put up one heck of a fight when hooked on light tackle. Other impressive California minnows with intriguing names include the splittail, blackfish, hardhead, hitch, and tui chub. These California specialties are but a small subset of approximately 250 cyprinid species native to North America.

A conceptual framework for my dissertation came to me in a flash of insight. Two famous paleontologists, Niles Eldredge and Stephen J. Gould, recently had hypothesized that virtually all evolution occurs during speciation events. In other words, almost all evolutionary change transpires at the nodes in evolutionary trees, where speciation takes place, rather than along the evolutionary branches (as assumed under traditional Darwinism). This controversial suggestion gave me an idea.

Suppose that two groups of organisms of similar evolutionary age had experienced greatly different rates of speciation in their histories. If the Eldredge-Gould model of "punctuated equilibrium" was correct, then living representatives of the rapidly speciating group should be more divergent from one another than members of the slowly speciating assemblage, because in their histories they have been through more rounds of speciation. However, under the more traditional view of "phyletic gradualism," in which evolution marches primarily to the beat of time, living members of the rapidly and slowly speciating groups should show similar levels of divergence, because both were of comparable evolutionary age.

With help, I formalized these intuitions into a set of mathematical models that yielded distinct empirical predictions for the accumulated amount of evolutionary divergence expected in rapidly versus slowly speciating assemblages under the competing hypotheses of punctuated equilibrium and phyletic gradualism. Such strongly dichotomous but viable predictions are rather rare in evolutionary biology and are to be cherished.

To test which of these expectations better matched empirical reality, I decided to compare genetic divergence patterns in the species-rich minnows (Cyprinidae) and the relatively species-poor sunfish (Centrarchidae). The former group includes about 250 living species in North America, the latter only about 30. Yet, both groups were represented by early fossils dating to the mid-Cenozoic. If the large disparity in numbers of living species reflected different rates of speciation, then a comparison of mean genetic distances in the minnows versus sunfish would likely be highly relevant to the ongoing debate between punctuated equilibrium and phyletic gradualism.

My first task was to collect the necessary specimens, and all I needed was someone to hold the other end of the seine. Members of the lab were eager to help. On several delightful trips, pairs of us roamed the state from the Central Valley lowlands to the Coast Ranges to the eastern deserts to the low and high Sierras. Each trip brought adventures and tribulations, but two stand out.

Marty Tracey was a brilliant and somewhat offbeat postdoc in the lab. He had a wonderful sense of humor and needed it. Marty had been applying for faculty positions for quite some time, but then as always the job market in evolutionary biology was somewhere between slim and nil. Out of frustration, he decided to have his many job-rejection letters leather bound, and he was working on volume three when I asked him to be my seining assistant for a collecting trip up the northern California coast. As luck would have it, Marty had been invited to interview at a small college in Medford, Oregon, near the California border. So, off we went to accomplish the joint missions.

We seined by day, camped by night, and generally made a grubby mess of ourselves on the three-day trip northward. We had planned to rent a motel room to clean up before the job interview, but we arrived in Medford late at night, and Marty had reverted to his usual cynical outlook on job prospects. Saying "to hell with it," he suggested that we camp under the stars and have a few celebratory beers to mark our successful collecting efforts. Marty went to the interview the next day in malodorous field clothes, unshaven, and a bit hung over. He said he had a fine time, but for some inexplicable reason wasn't offered the faculty position.

I was accompanied on another trip by the lab's chief technician, Lori Barr. One day, we were seining for hardheads in a stream of the Sierra foothills when an agitated rancher drove up, jumped out of his truck, and pointed a shotgun at our heads. He was convinced that we were cattle rustlers and that he had caught us red-handed. I think he might have shot us right on the spot had it not been for a woman's softening presence. Lori's plaintive tone was barely enough to calm him to the point where he agreed to peer into our ice chest to check our story, which to him seemed harebrained. With his boiling blood cooled to a mere simmer, he was finally convinced by the evidence: frozen fish, seines, rubber waders, university van, fish-collecting permits. Nonetheless, he ordered us out of the area in no uncertain terms, never again to return.

This wasn't my first field encounter with the barrel of a shotgun. A similar episode had occurred two years earlier when Mike Smith, Paul Ramsey, and I were seining a public creek in Alabama, near a sign that read "no huntin, no trespassin, no nothin." Paul and I were pulling the seine, and Mike, a few steps behind, was toting a plastic bag with our bluegill specimens. After one long seine pull, Paul and I emerged from the creek and looked up into the face of an irate farmer and the threatening twin barrels of his 12-gauge. We quickly turned to Mike, who we assumed would explain the situation, but to our chagrin, all we could see was his butt disappearing over the brow of a nearby hill. Paul and I talked our way out of the dilemma, but later, back at the van, we were anxious to hear Mike's explanation. He had seen the farmer coming and felt that his highest imperative was to save the fish collection. We took great solace in that sentiment from our boss.

I eventually gathered the minnow species, conducted the laboratory assays, and analyzed the data. From my days at SREL, I already had the necessary genetic information on sunfish. A comparison of the two data sets yielded results that at face value were consistent with expectations of phyletic gradualism and incompatible with punctuated equilibrium. Years later, our additional genetic studies of nearly seventy minnow species from the eastern United States further bolstered this conclusion for North American fishes. The published results were cited widely and in some circles hailed as important to the broader debate over major evolutionary patterns.

Nonetheless, there were important caveats to these studies, as duly noted in our original papers. Perhaps the evolutionary ages of minnows and sunfish were inadequately understood from fossil evidence. Perhaps different rates of extinction rather than speciation had been responsible for the divergent numbers of living species in the two groups. Perhaps the available taxonomies were misleading, such that the two groups weren't coherent evolutionary units to begin with. Perhaps the allozyme distances were inadequate measures of overall genetic differentiation. Perhaps all evolutionary change occurred during speciation events, but the magnitude of change per speciation was greater in sunfish than minnows. Finally, perhaps my interpretations were valid with respect to protein features (the subject of my assays) but not with respect to morphological features, which were the primary basis for the original debate between the phyletic gradualists and the proponents of punctuated equilibrium.

In any event, as an intellectual exercise my dissertation research educated me in ways that far outweighed the empirical findings alone. I came to appreciate even more the conceptual richness of evolutionary biology and to understand that innovative ideas or approaches often can be at least as edifying as technical findings per se. My appreciation grew for the illuminating role and also the sheer artistry of relevant mathematical or graphical theory. I also came to embrace the comparative method in evolutionary biology, wherein observations or ideas from disparate sources are juxtaposed to reveal phenomena that might not be apparent from the sources considered separately. Overall, I learned the merits of the strong-inference approach, whereby alternative hypotheses are erected and tested critically against empirical evidence. I also learned the importance of openly acknowledging the limitations of one's own data.

On a pragmatic front, I started to revel in the fine art of the written presentation and defense of scientific arguments. For example, I gained an appreciation of how to short-circuit criticism by actively seeking to identify and elaborate appropriate caveats to any stated conclusion. This necessity stems in part from the historical nature of evolutionary biology, where it is often impossible to know or to control all of the relevant variables that might have influenced a phenomenon under investigation. The same can be said of other historical sciences, such as geology and anthropology. Indeed, the full scope of conceivable results, or the range of theoretically viable possibilities, can often be of greater interest in evolutionary studies than the historically constrained outcome in any one instance. Furthermore, the exercise of identifying caveats to one's own work both sharpens the mind and protects it from falling into self-congratulatory ruts.

In general, Dr. Ayala's lab provided a fabulous educational environment. Francisco himself is a brilliant Renaissance man with eclectic interests ranging from the evolutionary genetic sciences to philosophy and religion. Trained early as a Jesuit priest in his native Spain, he maintained an active interest in the humanities, and indeed a significant fraction of his publications are in that area. He is also a gentleman scientist in the best sense of the phrase—courteous and professional to a fault, yet penetratingly insightful.

Another benefit of working in Francisco's lab was the opportunity to interact regularly with Doby. One day, he asked in his characteristically high, squeaky voice: "Potential Dr. Avise" (he always referred to us students that way), "would you care to chauffeur me on my last trip ever to Death Valley?" Doby seemed obsessed with thoughts of death and frequently prefaced statements with, "This will probably be my last" . . . new car, publication, trip, or whatever. I asked Francisco about it, and he told me not to worry—Doby had been saying such things for twenty years. In any event, I was delighted to have this travel opportunity for two reasons. First, I wanted to spend time in the field with this man, who had been a prime architect of evolutionary genetics in the twentieth century. Second, I wanted to see Death Valley and, in particular, its endangered pupfish. I couldn't have anticipated that these two aspects would intersect in an ironic event that occurred near the end of our five-day journey.

Off we drove from Davis, accompanied by Olga Pavlovsky (Doby's long-term research associate) and her husband Vadim, both nearly Doby's age. We toured Death Valley and the nearby Panamint and Amargosa Mountains, which brought back to Doby fond memories of his earlier fieldwork in the area. Finally, it came time to view the pupfish. About five species survive precariously as isolated relicts from ancestors who lived in lakes and rivers that filled Death Valley and adjacent hydrologic basins during the Pleistocene. From my readings, I was generally familiar with these fascinating creatures of desert springs and their striking adaptations for high temperature and salinity. Now, I wanted to see the pupfish in person.

Doby and I hiked along a section of Salt Creek that was the only native home in all the world to the endangered *Cyprinodon salinus*. Signs warned not to disturb the fish, but we caught only brief glimpses of a few individuals. At one point, as Doby crossed the stream to gain a better vantage, he teetered briefly on a wobbly rock. As he jumped to the far bank, the crushed bodies of three small pupfish floated downstream. I didn't mention the incident to Doby, and it certainly was of no real consequence to the species, but that image of the great evolutionist, ironically juxtaposed against the deceased pupfish in that beautiful desert stream, left a lasting impression. When I revisited Death Valley twenty years later, I was delighted to see that pupfish were present by the thousands and that the Park Service had constructed a pedestrian boardwalk along Salt Creek.

During that trip, Doby commented that these might be the last of such field delights for him. He died suddenly about a year later, in December 1975. One fascinating aspect of my experience in Davis was Doby's conviction that life had some ultimate meaning. This seemed at face value to contradict my own developing understanding of evolution. Natural selection, mutation, genetic drift, Mendelian inheritance, and other such mechanistic operations give every objective indication of being merely nonsentient, amoral forces of nature. Yet, Doby seemed convinced that any evolutionary process culminating in human beings must be purposive in some ultimate sense. Doby wanted no part of a "Devil's vaudeville" in which humankind ended in ultimate oblivion or in which the universe became "spiritless."

Two books that Doby wrote for a general audience, *The Biological Basis of Human Freedom* and *The Biology of Ultimate Concern,* are evidence of his interest in broader philosophical issues. Both Doby and Francisco were great enthusiasts of the French philosopher Pierre Teilhard de Chardin, a paleontologist and Jesuit priest, who saw mankind as the culmination of evolution and who tried to develop a spiritual ideology from that scientific base. Because Teilhard de Chardin was admired by two scientists whom I so respected, I began to read his works too. However, I got precious little from the effort. Perhaps this merely reflects my intellectual shallowness, but then and now, Teilhard de Chardin's writings are far too obtuse, allegorical, and mystical for my taste.

I completed my Ph.D. at Davis in 1975, also the final year of U.S. military involvement in the Vietnam War. I had begun junior high school fifteen years earlier, in 1959, one year after the first of 58,000 Americans lost his life in combat. Their average age was twenty-three. Many times that number of Vietnamese were killed, and countless more young people on both sides were physically injured or mentally scarred for life. To what avail was this prolonged horror? Apart from some possible historical lessons that we may (or may not) have learned, I have no idea.

I had been spared the trauma of fighting in Vietnam, and indeed the war had accidentally nudged me toward an unanticipated career path in science. I had pledged in my CO document to try to do something positive with my life, something that would benefit rather than harm the planet. Would I now be able to meet that commitment?

8 Getting Started

A career in academic science is unlike nearly any other. In effect, you are hired by a nonprofit organization for intellectual services. You don't exactly have a boss, although you are assigned teaching responsibilities and must report periodically on these and all research activities to university administrators. By far the greatest attraction of academic life (apart from seldom having to wear a tie) is that you have the freedom and autonomy to pursue virtually any serious line of inquiry that you wish. Essentially the only requirement is that you remain visible and productive in your chosen field. Basically, this means publish or perish.

Apart from salary, space, and a modest setup, which the university provides, it's strictly up to you to supply the wherewithal to conduct your research. This means that you must devise compelling scientific questions, write formal grant proposals that "sell" your ideas, typically to a federal agency or foundation, use these funded awards to set up and run your laboratory, publish your lab's findings in scientific journals or books, and otherwise disseminate the results via lectures at home and abroad. And that's only the research side of faculty life! In other words, a successful academic scientist must wear many hats. I was quite naive about all of this when I began my faculty career at the University of Georgia in 1975. On the other hand, my innocence probably freed me from preconceived notions and thereby enabled me to wander a less orthodox scientific path.

In the mid-1970s, few postdoctoral appointments were available in evolutionary biology because the field was small and poorly funded. Thus, as I neared graduation from Davis in 1975, I submitted job applications to all of the twenty or so colleges and universities advertising faculty positions for which I might be suited, however remotely. These ranged from marine biologist to desert ecologist, and from population geneticist to molecular systematist to museum curator. I made the short list for seven positions and soon was interviewed for each. My favorite interview was at the Hawaii Institute of Marine Biology. Alas, no job offer was forthcoming. Fortunately, bids were eventually extended from the University of Georgia's Zoology Department and the University of Houston's marine station at Galveston. The former would be a typical academic position, and the latter entailed applied genetic research on commercial shrimp in a small field facility.

I agonized over the decision. How could I pass up the extraordinary opportunities afforded by Georgia's high-profile research university, despite the inevitable pressures that would accompany an academic career in such a large and competitive environment? Wouldn't my day-to-day life be more enjoyable in the comfortable setting of a small field laboratory on the Texas shore? My answer to the second question was yes, ergo the dilemma. Being young and ambitious, I chose the University of Georgia, but I've often wondered how my life might have unfolded had the Galveston job been the choice instead.

In the summer of 1975, I again drove cross-country in my Mustang convertible, all of my worldly possessions crammed inside. As I entered the Southeast, the pine trees, mockingbirds, and rolling hills of red clay felt comfortably familiar, like old friends. The move was a homecoming in terms of natural history and landscape, but in other respects I would be entering personally uncharted territory. I rented a small apartment in Athens and began to set up shop at the university. My nine-month salary was $14,000, nearly six times my graduate-student income. Although suddenly wealthy by prior standards, my personal lifestyle changed little with faculty status. When Joan and I married and purchased a modest home a few years later, the extra dollars merely converted from apartment rentals to mortgage payments, and any spare time transmuted from fishing to lawn mowing.

The university's offer also included $10,000 for furnishing my research laboratory. The space I moved into had previously been occupied by another population geneticist, so I supposed that some used equipment might remain available as well. This proved not to be the case. Everything left behind (including electrical outlet strips on the walls) had been picked clean by departmental gremlins. There was one blessed exception. On a dusty shelf sat a large box of Whatman #3 filter papers neatly cut into 5 by 8 mm rectangles. These became the wicks (to soak up tissue samples) that I would employ by the tens of thousands over the next decade in countless protein-electrophoretic gels.

My primary fears about the move to Georgia were predictable. Could I lecture effectively on a daily basis to large undergraduate classes? Could I convey the technical material required of graduate courses? Would committee duties be bearable? Could I devise research projects and write grant proposals that would attract grant support? Would I publish a sufficient volume and quality of research to earn my keep? Most of these fears proved to be well founded. Lecturing always has been a trying endeavor for me. Committee work is a vast and mostly unproductive time sink. With regard to funding, my first several grant proposals were flatly rejected. Thus, apart from the paltry setup dollars, my laboratory went unsupported for the first three years.

Throughout my career in evolutionary biology, an endless frustration has been the uphill battle for meager research funds. Unlike medical and agricultural sciences, which are *relatively* flush, most basic biological research in the United States is chronically undersupported. For example, the Biological Sciences program at the National Science Foundation (NSF), the primary granting outlet for most of this country's nonapplied biological research, even today has an annual budget of little more than $320 million. This is only one-fortieth the annual budget of the National Institutes of Health, a federal agency supporting medical research. It also happens to be less than 25 percent of the cost of one B-2 Stealth bomber. Furthermore, most of these NSF dollars go to molecular biology, cell biology, and neurobiology, with only loose change left over for the other life sciences, including evolution, systematics, ecology, behavior, biodiversity analysis, and natural history.

This means that the competition for funds is fierce, and about 85 percent of new grant proposals typically are rejected by NSF (despite the months of thought and preparation that go into each). Because the rejection of grant proposals is the rule rather than an exception, a career in academic science is not for the thin-skinned. Critical evaluations of one's best efforts are a normal part of the scene. Administrators and peer-reviewers are just doing their job, which is to search hard for the empirical or interpretive shortcomings of one's research, rather than to lavish praise on its possible merits. The unrelenting harshness of peer review, particularly in a tight funding climate, is entirely proper and crucial for scientific advance, but it can be extremely disheartening.

In truth, I have never been very good at attracting grant support. My laboratory at Georgia has now been in operation for more than twenty-five years, during which time I have brought in a grand total of about $825,000 in grant funds, mostly from NSF. This may seem like a lot of money, but it is a very modest sum in comparison to what is consumed by most molecular genetics labs. During that time, my students and I have published on average one scientific paper every six weeks, most of them appearing in high-profile journals. This works out to about $3,900 per publication, a mere pittance by the normal research standards. (I'll bet that the unit cost of a scientific paper in medicine or epidemiology, for example, averages ten to one hundred times higher, although some researchers might suggest that this is like comparing apples and oranges.) I have finally come to interpret my lab's low funding, and its associated high research output *per dollar*, as a source of pride rather than shame, although university administrators, preoccupied with their bottom line of total grant dollars received, utterly fail to follow this logic. In any event, as a practicing scientist, I feel deeply privileged to be in academic research and extremely grateful for *any* societal patronage of the life sciences.

Fortunately, I've usually had far less difficulty with publishing, and this was true even in my earliest years at Georgia. Here's why. Several months before graduating from Davis, I "secretly" completed but did not submit my bound dissertation. I used the extra time, apart from the job interviews, to learn new techniques and push forward laboratory work on several other research projects. Like any well-trained Swede from the Upper Peninsula, I was stockpiling data for the harder times likely to come.

A postdoctoral friend of mine at Davis, John Gold, was an expert on chromosomal (karyotypic) analysis, and I asked him to teach me the tricks of the trade. Together, John and I surveyed the chromosomal makeup of various California fishes. We stumbled upon a triploid minnow (possessing three chromosome sets rather than the usual two) and discovered some rare instances in which a few of a fish's cells carried up to twelve sets of chromosomes each. We also compared karyotypes among the California minnows species. I took these and other data with me to Georgia and there wrote them up into several publications. It was not spectacular science, but it did help me remain productive at a time when academic tenure still hung in the balance.

Another advantage I had in those early faculty years stemmed from my prior association with SREL, which happens to be affiliated with the University of Georgia. This ecology facility has always been well funded by federal grants supporting its mission-oriented activities, and Mike Smith was kind enough to let me return to work in his lab during the summers. I rented an apartment in the town of New Ellenton and commuted daily to SREL.

Although many of my initial projects at Georgia and SREL involved continued genetic analysis of population structure in fishes, I ventured further afield as well. For example, one year I worked with a faculty colleague, Sue Duvall, on monkeys housed at the Yerkes Primate Center, near Atlanta. Several macaque species resided there, including the lion-tailed, pig-tailed, crab-eating, and black-stumptail. Notwithstanding their morphological and behavioral differences, these species interbreed occasionally, and several hybrid animals had been born at Yerkes. We took blood samples from these animals for genetic analysis to address a hypothesis popular at the time concerning patterns of gene regulation.

The time and bodily location that a given protein-coding gene is activated during an organism's development depends in part upon other genes that constitute the regulatory apparatus of a genome. Different genes of a species have evolved to interact properly with each other, but there is no reason to suppose that regulatory and structural genes of two separate species that evolved independently for thousands or millions of years should interact properly when thrown together in hybrid animals. Although no method then existed to assay regulatory genes directly, one indirect approach was to examine patterns of structural gene expression in hybrids. There, the divergent maternal and paternal genomes might fail to interact properly, perhaps yielding instances of "allelic repression," which might serve as a quantifiable gauge of regulatory breakdown.

Dr. Duvall and I wanted to test whether protein-electrophoretic assays might document such regulatory incompatibility. Specifically, did the fraction of protein-coding genes whose expression was disrupted in hybrids correlate with the magnitude of genetic distance between the parental species? We conducted the assays, and the results were clear. Both the maternal and paternal alleles were fully expressed in all hybrid macaques, regardless of how different the parental species were genetically. Either the breakdown in gene regulation was limited, or our procedures of protein assay were inadequate measures of the phenomenon.

As a follow-up project on primates, Sue and I planned to study an endearing group of animals endemic to Madagascar. About twenty species of lemurs exist, ranging from the mouse-sized dwarf lemur to cuddly sifakas, the size of a medium dog. We were especially interested in phylogenetic relationships within the assemblage and how and when such a diverse array of species might have arisen on that large island off Africa's east coast. Perhaps the study of these animals might reveal aspects of speciational processes analogous in some ways to those that fostered the "flocks" of fish species in Lake Lanao and in some other closed bodies of water around the world.

To get blood samples, Sue and I made preliminary inquiries and visited the Duke University lemur facility in North Carolina. However, the project soon came to a tragic end. In an inexplicable burst of rage, a disgruntled graduate student in Georgia's Zoology Department shot and killed Sue, and then turned the gun on himself in a horrible murder-suicide. This devastating loss of a close friend and scientific colleague left me with no desire to continue the lemur project or, indeed, to work on other primate species.

One unexpected challenge in my transition to faculty status proved greater than any other: supervising graduate students. Previously in life, I had endeavored primarily on my own behalf, taking courses for degrees and personal edification, living off laboratory resources and intellectual ideas supplied mostly by my academic advisors, and in general focusing on Number One. Like any student, I took this modus operandi for granted. Now, for the first time, I felt a burden of responsibility for the career fate of others. In that and other regards, having graduate students is not unlike having children. Indeed, it is customary in academia to refer to scientific pedigrees. For example, Dr. Ayala is my academic father, Doby was my grandfather, and John McDonald is one of my many academic siblings.

My first graduate advisee was Mike Douglas (who many years later became director of the Vertebrate Museum at Arizona State University and editor of the journal *Copeia*). Mike had approached me after taking my advanced course in evolution. Here was a sobering new challenge, made doubly worrisome because no grants to support my lab's operation had been forthcoming, and I had nearly exhausted my start-up funds. Also, neither Mike nor I was competent in formal mathematical theory, so that inexpensive research avenue was closed. Thus, we had to devise an empirical dissertation project that involved little or no monetary outlay. Fortunately, I had an idea.

Steve Gould had read the publications resulting from my own dissertation regarding the ongoing debate between phyletic gradualism and punctuated equilibrium. He applauded in print the conceptual novelty of my approach but countered that the debate really was about patterns of morphological rather than protein evolution. Indeed, under the neutral-mutation theory popular at the time, molecules might be expected to evolve more or less in a time-dependent rather than speciation-dependent fashion. Thus, he suggested, my allozyme results merely reflected neutral evolution and, hence, were not directly germane to the punctuated equilibrium issue.

In response, it occurred to me that the theoretical models predicting relative levels of genetic divergence in species-rich versus species-poor evolutionary groups should also apply, with suitable modification, to morphology. Thus, perhaps we could retest the punctuated equilibrium hypothesis more directly by comparing levels of morphological divergence in minnows and sunfish. By measuring morphological traits on these fish and statistically analyzing the data, Mike could fashion a worthwhile thesis at practically no monetary cost, save a set of $50 calipers. I vividly remember asking the departmental chairman whether his office might please buy us the calipers, with my promise that a successful Ph.D. graduate would be the end result. "OK," responded the chairman, not being one to pass up such a good deal!

So, Mike spent most of his graduate tenure in virtual slave labor, gathering and analyzing morphological data from more than fifty species. He counted fin rays and scales, measured swim-bladder volumes, and in general poked and probed into every measurable crevice of each fish's body. The results boiled down statistically to a striking conclusion. The living minnows as a group were similar in overall morphological diversity to the sunfish, a result generally inconsistent with the predictions of punctuated equilibrium. There were again caveats to this conclusion, so the study was less than definitive. Nonetheless, the exercise was intellectually challenging and resulted in provocative publications as well as a Ph.D. degree.

I have supervised many graduate students over the years, producing nineteen Ph.D.s and two M.S.s at latest count. All have gone on to successful biological careers, often but not invariably in academia. I realize that this number far exceeds my "rightful" allotment of students. In a field such as evolutionary biology, which has grown only slowly over the years, on average only about one or two protégés per generation can succeed each mentor. Indeed, this reasoning was one basis for a recent report by the National Research Council (the advisory arm of the National Academy of Sciences) admonishing that too many advanced biology degrees are being issued in the United States relative to the country's needs.

I take strong issue with that conclusion, particularly with respect to the environmental and genetic sciences. At no time in human history have societies been *more* in need of trained ecologists and geneticists. We are in a time of unprecedented environmental crises on both local and international scales, with problems ranging from high rates of species extinction to global warming. Most marine fisheries have collapsed economically or are unsustainable at current rates of harvest; biota-rich wetlands are being drained and tropical rain forests cleared at alarming rates; underground aquifers and fossil-fuel reserves that took geological epochs to fill are being depleted in our lifetimes; surface and ground waters are polluted with agricultural and industrial runoff; greenhouse gases threaten the earth's climatic regimes, upon which our agriculture and even settlement patterns depend. Such lists can go on and on. Who will be capable of appreciating and addressing such bio-environmental challenges?

On another major biological front—genetics—current challenges and opportunities likewise are unprecedented. Technological breakthroughs in genetic engineering permit such laboratory wizardry as organismal cloning, directed mutagenesis, and artificial swaps of genes among life forms as different as bacteria and humans. Who will be able to harness these awesome powers for human and environmental benefit, and who will wisely counsel societal attempts to avoid the potential abuses and dangers? In one of the keynote empirical achievements in the history of science, the human genome has essentially been sequenced in its entirety (all three billion bits of information), providing the raw coded text of the genetic encyclopedia of our species. Who will be competent to read and interpret this magnificent work of evolution? Who can put that information to use in developing treatments and cures that the medical profession now may scarcely imagine? From a sound scientific base, who will consider or give advice on the daunting ethical and social ramifications?

My argument is that many more, not fewer, well-trained biologists are needed now and in the foreseeable future to address societies' *genuine* needs. Furthermore, students trained in the biological and environmental sciences typically have problem-solving talents and communication skills, also well suited for a wide range of nontraditional career paths. Several of my own students illustrate this point. For example, John Patton has operated an ecological-genetic consulting company, and Lou Kessler is scientific advisor to a firm of patent attorneys in Washington, D.C.

More generally, why shouldn't knowledgeable biologists and other scientists frequently be sought out for political office and other positions of authority traditionally reserved for businessmen and lawyers? Scientifically literate leaders in business and government might be far more likely than conventionally trained bureaucrats to spawn policy initiatives that could foster economic or societal well-being without compromising the health of our environmental life-support systems. Furthermore, any initial policy successes should promote a positive feedback loop, leading to new career opportunities for additional scientists in the public and private sectors.

Each graduate student has unique talents and proclivities, and affords singular challenges. Students have come to my laboratory with backgrounds ranging from field ecology to molecular genetics to statistics and math. Their native dispositions have ranged from abstract conceptualization to meticulous empiricism. Thus, a major challenge for the faculty advisor is to capitalize upon each student's inherent strengths, to lessen or overcome his or her weaknesses, and to promote a working environment that fosters collaborations where complementary abilities can be assembled to mutual benefit. Finally, I have come to relish rather than lament the great diversity of ways in which students' minds seem to work in solving the many challenges of scientific research.

In late 1978, I finally won my first substantial federal research grant from the National Science Foundation. Shortly thereafter, I was promoted and awarded academic tenure. For the first time, this Georgia Bulldog really began to see himself as a true academician for the longer haul.

9 Class Aves

It takes considerable effort to design and implement a university course. Under the quarter system at Georgia, nearly fifty different hour-long lectures must be prepared and delivered over a ten-week session. If laboratory work is involved, the syllabus includes thirty lectures and twenty lab meetings. A typical teaching load in the biological sciences at Georgia involves 1.5 courses per academic year, or the equivalent of about seventy-five lectures. In my early years (and continuing today), I rotated through several different classes in genetics, evolutionary biology, and molecular evolution.

One day, my department chairman asked me to teach ornithology as well. He knew of my birding interests, and my surname had a decidedly avian ring to it. I didn't have tenure at the time, so naturally I agreed. It was a wise "decision." Over the years, ornithology has been fun to teach, and it also has enabled me to keep one foot planted in natural history, even as the other was often sucked into the seductive quicksand of molecular genetics.

I had no formal training in ornithology, so I was soon poring over avian textbooks and journals trying to master bird anatomy, physiology, ecology, paleontology, systematics, and conservation. The only topic I didn't have to worry about was field identification, because I knew the birds well. The material is organized into three lectures per week, plus lab sessions consisting of indoor exercises and local field trips. The trips emphasize natural history, behavior, and the field identification of about 240 local species, including 70 by song. There are also weekend excursions to the North Georgia mountains and to barrier islands on the state's coast.

The trip to the mountains targets many avian species that normally nest across the northern United States and Canada but also occur at high elevations in the southern Appalachians. Georgia's highest peak, Brasstown Bald, at nearly 5,000 feet, displays a climate more like that of New England than the Southland it rises above. Searching the streams, hemlock forests, and rhododendron thickets along its flanks, my students and I encounter delightful species otherwise characteristic of northern climes. These include the raven, veery, rose-breasted grosbeak, solitary vireo, and several lovely warblers who first reveal their presence by characteristic songs: the buzzy phrases of the black-throated green and black-throated blue, the monotonic trill of the worm-eating warbler, the clear-toned renditions of the Canada and chestnut-sided, and the emphatic "teacher-teacher" of the ovenbird.

The coastal trip is even more enjoyable. Georgia boasts some of the last remaining unspoiled stretches of shoreline in the eastern United States. Situated at the margin of a broad embayment known as the South Atlantic Bight, about a dozen barrier islands protect broad swaths of tidal marshlands on their leeward side. These beautiful islands, just a mile or two wide, are geologically active spits of sand continually reshaped by ocean currents. Clothed in mature forests of live oaks and pines, dotted with freshwater ponds, and flanked by beaches and marshes, they teem with wildlife. Most of the islands have been in single-family estates or under federal or state stewardship, accounting for their undeveloped status.

You do not just drive to the beach in Georgia, you drive to the coast, where you are greeted by an expanse of *Spartina* marsh reminiscent of a tall-grass prairie. A labyrinth of tidal creeks meanders through this swaying sea of green and brown, bathing and flushing the slough and nourishing its rich biota. The wider creeks also serve as transportation corridors for shrimp trawlers and crab potters to reach fertile fishing grounds in the estuaries and ocean inlets, and for small passenger ferries to transport local residents to the barrier islands. Only after reaching an island and trekking across to its outer side do you finally encounter sandy beaches facing an open ocean.

Like inland prairies, salt marshes are an acquired taste. The initial impression is one of a monotonous, gnat-infested savanna, but the marsh's treasures are subtle and sublime. Pleasant hours can be lazed away on any rickety dock along a tidal creek, listening to the whistles and clucks of boat-tailed grackles, the sewing-machine songs of marsh wrens, and the clicking calls of clapper rails. From that same vantage, you can monitor the activities of osprey, pelicans, least bitterns, and several species of herons, egrets, gulls, terns, and sandpipers.

You can also jig for blue crabs or cast a line for sea trout, redfish, and croaker.

You can find amusement in the fiddler crabs scrambling about the mudflats or in mullet and killifish playing in the waters below. Or, you may prefer simply to inhale the lusty aroma of salt marsh humus, admire the moving panorama of grass and water, and ponder vivid cloud formations against an open sky. As such impressions invade the senses, an appreciation for marshland habitat can grow into a comfortable romance that grips the soul.

My favorite class field trip is to Sapelo Island. We depart Athens late on a Friday morning for the five-hour drive to the coast in two crowded university vans. The bird list, which will grow to include about 130 species by trip's end, begins immediately with students calling out sightings of red-tailed hawk, eastern bluebird, loggerhead shrike, and American kestrel as we proceed southeast along country roads. We can't dally because the ferry leaves the dock promptly at 5 P.M., with or without us. Arriving with a half-hour to spare, we transfer to the boat for an hour-long ride to Sapelo. For many students, this is their first exposure to a salt-marsh environment.

The university's field station on Sapelo is a repository for the state's most decrepit vehicles. Arriving at the Sapelo dock, we pile into a one-ton truck fitted with wooden benches on an open flatbed behind the rusty cab. Although this ancient jalopy lacks many accouterments, such as a muffler, treaded tires, door latches, paint, or operational windows, it is suited perfectly to our needs. We are here to ramble the island and savor its ambience. When wrestling the truck along sandy roads, I have to yell out "Duck!" at appropriate times so that students in back aren't injured as we press forward through tangled vines or under the limbs of moss-draped oaks.

Our dormitory is equally dilapidated, yet lovable, with a moldy aroma and squeaky wooden floors and staircases. From its antique kitchen will emerge home-cooked meals prepared by two warmhearted townswomen from the nearby settlement of Hog Hammock. The rest of our stay, from predawn marsh treks to postdusk owl walks, becomes a frenetic blur of birding and nature activity.

At low tide, we roam the outer beaches for avian specialties such as black skimmers and sandwich terns resting on the sand flats; American oystercatchers, willets, and Wilson's plovers nesting on the back dunes; and northern gannets diving offshore. We hike the inlet margins, checking for common loons, red-breasted mergansers, and scoters on the estuary and numerous species of wading birds along the shoreline. We scan the skies for species ranging from swallows to bald eagles and wood storks, and we search freshwater ponds for ducks, coots, rails, moorhens, and purple gallinules.

We comb the woods for wild turkeys, hawks, wrens, woodpeckers, nuthatches, flycatchers, tanagers, orioles, thrushes, and the outrageously colored painted bunting, arguably the most striking bird in North America with its patchwork plumage of purple, scarlet, and green. We visit a heron rookery at the north end of the island to see snowy and common egrets, great blue, little blue, tricolored, and green herons, and black-crowned and yellow-crowned night herons. We tramp a mile-long dike through the marsh to an old lighthouse where barn owls roost, searching shrubs along the way for migrant warblers, vireos, and thrushes that arrived during the night. By the time we return home to Athens late on Sunday, we are exhausted yet profoundly satisfied.

To bolster my professional credentials in ornithology, I decided to conduct some formal genetic research on the class Aves. A first step was to obtain tissue samples from species of interest. Waterfowl seemed one logical choice because game birds can be collected with relative ease, and several species had uncertain phylogenetic affinities. With state and federal permits in hand, I and a student, Lou Kessler, were ready to begin a genetic survey of ducks. We would collect some of our samples at the Welder Wildlife Research Station near Corpus Christi, Texas. This site is a popular winter home for waterfowl from both the eastern and western United States, and it had suitable lab facilities for the initial tissue workups.

Lou and I were new to duck hunting and not very good shots. After many unsuccessful shotgun blasts during our first morning in the field, I finally bagged a lesser scaup. When we retrieved the body, Lou noticed a metal band on its leg inscribed with the words, AVISE: WRITE WASHINGTON, D.C., plus a long number. Lou did not realize that these tags are used routinely by the U.S. Fish and Wildlife Service to monitor duck migration. The word "avise," roughly translated "advise," is meant to encourage hunters to notify wildlife officials in Washington of the site (often in Latin America) where a leg band was recovered. Of course, I didn't tell Lou this. With a straight face, I explained that my collecting permit specified that we were under strict government orders not to kill any ducks except those labeled with my name. So, I told Lou to scrutinize carefully, as I had done, each duck that flew overhead to make sure its band was properly inscribed before pulling the trigger. Poor Lou was left incredulous at the whole situation.

Fortunately, the abundance of waterfowl at Welder gave us many more shooting opportunities, and we completed the collections within a week. Later, back in Athens, the samples were analyzed genetically and the data interpreted with regard to phylogeny. Several interesting findings emerged, such as the fact that despite external appearances, the cinnamon teal is related more closely to the shoveler duck than it is to the green-winged teal. This result seemed inconsistent with the body sizes (and the common names) of these species, but it made sense in terms of a striking morphological feature, a big blue patch on the forewing, shared by the shoveler and cinnamon teal but not by the green-winged teal.

Such findings raised two broader points that seem obvious now but were less so then: the overall appearance of an animal sometimes can be a misleading guide to its historical (phylogenetic) affinities, and the evolutionary histories of particular traits often can be deduced by mapping them onto molecular-based estimates of phylogeny. Thus, relational aspects between genetic and morphological data often are of greater interest than either set of information considered alone. Molecular and organismal systematics in those days were often perceived as being at odds, but I came to see their distinctive data bases as mutually informative. Indeed, many years later, "phylogenetic character mapping" became standard practice in systematics. This method entails placing various structural, behavioral, or physiological traits onto a genetic-based estimate of phylogeny to reconstruct the evolutionary history of alternative forms of these features.

I wanted to extend our phylogenetic analyses to other avian groups, but obtaining tissue samples presented ethical as well as logistic difficulties. Even if collection permits could be obtained and the birds encountered in suitable numbers, how could I countenance killing adorable creatures like vireos and thrushes, let alone the wondrous North American warblers, which had occupied a special place in my heart since childhood? I was relieved but also saddened when this dilemma soon resolved itself through contacts I made at a small biological station in northern Florida.

The Tall Timbers Research Station is a privately funded facility devoted to the science of fire ecology. Situated north of Tallahassee, its primary mission is to understand fire's crucial role in perpetuating the ecological community of wiregrass and longleaf pine that formerly dominated the southeastern coastal plain. For thousands of years, this ecosystem was sustained by fires (ignited by lightning or Native Americans) that periodically cleared the forest understory yet spared the mature longleafs. These widespread forests supported many specially adapted species, including the indigo snake, red-cockaded woodpecker, Bachman's sparrow, and gopher tortoise, which are listed as threatened or endangered in today's forest remnants.

Beginning in the last century, the lumber industry clear-cut most of these ancient longleaf forests and replaced them with vast monocultural plantations of faster-growing pines, arranged in meticulously straight rows and harvested on a twenty-year cycle. Notwithstanding disingenuous "green ads" to the contrary, most of these dreary replacement forests are biological deserts. Such is the price we pay for our magazines and thick daily newspapers made from the pulp of these scrubby crops. Ironically, daily reminders of the magnificent forests that once were can be found on Georgia's license plates, which depict a bobwhite quail flying from a clump of wiregrass beneath bows of longleaf pine.

Tall Timbers happens to be located across the road from the 1,000-foot-high WCTV transmitting tower, surrounded by thirty-four acres of closely mowed grass. This television tower, supported by wires that radiate to ground points up to 800 feet from its base, killed on average about 1,600 birds per year, mostly during spring and fall migrations. This fact had been documented, because each morning for more than twenty-five years, Tall Timbers' scientists scoured the WCTV grounds for birds that had died during the night after colliding with the lighted tower itself or its supporting cables. These avian remains were carefully identified to species, aged, sexed, and stored in freezers by the Tall Timbers staff.

Over the years, more than 42,000 kills had been registered at the WCTV tower, and this merely gives a hint of the problem because, before morning light, many of the carcasses are eaten by great-horned owls, feral cats and hogs, foxes, bobcats, skunks, raccoons, opossums, and other scavengers. To document the activities of these nocturnal freeloaders, scientists one evening scattered sixty marked avian corpses across the WCTV tower grounds. By early morning, all had disappeared!

Avian mortality through collisions with man-made structures is, in general, a major ecological problem. In my readings, I had seen references to the large numbers of birds killed by striking home windows, cars, and other artificial structures. Even more surprising was how many nocturnal migrants die after hitting skyscrapers, lighthouses, and other high-rise structures. For example, on September 30, 1973, 2,000 birds died from colliding with a smokestack in Cheshire, Ohio. The worst recorded disaster of this sort occurred on the nights of September 18 and 19, 1963, when an estimated 30,000 birds met their death at a single TV tower in Eau Claire, Wisconsin. Thus, the situation at WCTV is far from unique.

During migratory periods, at least some kills occurred every night at the WCTV tower, but the toll increased markedly with overcast conditions associated with a cold front. Then, the nocturnal migrants fly lower and are disoriented by the tower's lights, crashing into the structure or its supporting wires. The worst recorded slaughter occurred on the night of October 9, 1955, when 1,988 dead and dying birds of 62 species were tallied. These were among an estimated 4,000 to 7,000 corpses littering the ground that morning, the rest being lost to scavengers or partial decomposition before they could be cataloged by the beleaguered Tall Timbers' staff.

On several visits to Tall Timbers, I accompanied ornithologist Bobby Crawford on his daily rounds beneath the tower, picking up kills from the prior night and adding these to specimens gleaned from the laboratory freezers. Even against the closely mowed tower grounds, the small avian corpses were difficult to spot. Presumably, many of this nation's thousands of TV and radio towers likewise kill many birds during migration, yet the events typically go unrecognized, because most of the bodies remain undetected.

Each corpse under the television tower was a small tragedy to me. Particularly in the springtime, I couldn't examine the lifeless body of a blackpoll or blackburnian warbler without thinking of what had been lost: a gorgeous creature who, just a few days earlier, had left its winter residence in Venezuela to fly nonstop across more than a thousand miles of open ocean; a small animal whose feats of navigational skill, exertion, and daring would put the most heroic of such human efforts to shame; a tiny, ebullient bird in the prime of its life, driven by an age-old instinct to return to its natal home in Canada to perpetuate its kind. All of that was brought suddenly to an end by a cold, thin cable of steel.

In making the best of a sad situation, researchers over the years have put these Tall Timbers' tower-kill birds to beneficial use. From the tallies themselves, information has been gained on migration patterns. For example, some species must migrate in large flights, as suggested by the 2,000 palm warblers killed on October 8, 1955; the 54 summer tanagers found on October 5, 1957; or the 104 gray-cheeked thrushes on May 2, 1964. Some species use different migration routes in the spring and fall, as evidenced by the fact that all 466 bay-breasted and 483 chestnut-sided warblers cumulatively recovered at the WCTV tower died in the autumn, whereas 158 of the 163 blackpoll warblers were killed in the spring. In some species, different age classes may concentrate along different migration paths, as suggested by the fact that adults constituted only 29 percent of the 2,410 common yellowthroats killed at the WCTV tower, whereas they represented 64 percent of the kills at a similar tower far to the east in peninsular Florida.

Researchers also have utilized the frozen specimens in studies ranging from seasonal patterns of fat deposition to levels of pesticide residues in the birds' bodies. This tissue repository became an invaluable resource for my laboratory as well. Soon, I was making overnight trips to Tall Timbers to sort and retrieve recently obtained samples of thrushes, vireos, warblers, and sparrows for genetic analyses conducted back in Athens by me and graduate students John Patton and Chip Aquadro. We wanted to examine genetic relationships within these particular taxonomic groups of birds and compare the results to comparable genetic data from other creatures, such as mammals and fishes.

The business of systematics—generating phylogenetic trees and erecting classifications from them—traditionally has been something of an art, in which practitioners employ idiosyncratic kinds of data and taxonomic criteria unique to each animal or plant assemblage. For example, an avian systematist might search for phylogenetic clues in the details of a bird's leg musculature, beak, or feather alignment; an ichthyologist scrutinizes variation in the gill apparatus, fins, and swim bladder; and a mammalogist scores pelage features and tooth arrangement. By analyzing such data, phylogenetically related organisms can be identified *within* each group, and biological classifications erected accordingly. For example, all woodpeckers share a chisel-like bill, stiff tail feathers, and several other morphological and behavioral features indicative of close evolutionary ties, and accordingly they have been placed into one taxonomic family—Picidae. However, because the taxonomic features scored in woodpeckers and other birds differ from the types of features that can be scored in mammals and fishes, there was no common basis for assessing whether the family Picidae, for example, was equivalent in any meaningful way to a taxonomic family of fishes or mammals, much less of fruit flies or starfish.

Biological macromolecules afford quite a different perspective. Many cellular capabilities such as carbohydrate metabolism are nearly universal to life. So too are the genes that encode enzymatic and structural components of these biochemical operations. In principle, molecular assays could tap phylogenetic information from hundreds or thousands of analogous and often homologous genes or their protein products in virtually any species. Thus, appropriate molecular-genetic data can provide common denominators for assessing phylogeny in a comparative vein. Ultimately, the entire tree of life could be revealed. By integrating such information with the fossil record and other traditional sources of phylogenetic data, taxonomists should someday even be able to summarize this tree into a universally standardized temporal scheme for biological classification. Such an achievement would qualify as an empirical milestone in the history of biology.

The genetic data that we gathered from the Tall Timbers' birds yielded several phylogenetic insights for each taxonomic group. Of greater interest, however, was an emerging comparative picture. Prior research had given me familiarity with protein-based estimates of genetic distance between closely related species of mice, sunfish, minnows, and several other vertebrate and invertebrate groups. Interpreted against this backdrop, the genetic differences between bird species often proved to be rather small. In other words, at a given level of taxonomic recognition such as genus or family, birds often showed significantly less differentiation in allozymes than did many nonavian creatures of identical taxonomic rank. Similar conclusions later would be drawn from more refined protein assays, as well as DNA-level comparisons. It was as if avian taxa on average were about one taxonomic level out of step with much of the biological world.

Such disparities could have arisen if avian systematists had been taxonomic "splitters," whereas nonavian systematists had been relative "lumpers." Our early molecular studies on birds further opened my mind to the novel research prospects afforded by *comparative* assessments of molecular and morphological information. They also opened a Pandora's box of exciting issues related to the concept of common yardsticks in evolution, issues that I would return to frequently throughout my career.

After twenty-five years, I continue to include avian taxa as subjects of genetic research in my laboratory. This personal interest in birds also keeps me focused on ecological issues. Wild birds, like miners' canaries, are sensitive indicators of environmental problems. Avian diversity and abundance are excellent barometers of the ecological health of a nation and its attitude toward nature. In nearly all corners of the globe, native avifaunas give the distinct impression of a natural world under siege, as illustrated by the following vignettes.

Consider some of the Mediterranean countries. My initial clue to the sorry state of biodiversity in this region came from comments made by two visiting researchers to my lab, Lorenzo Zane and Donatella Crosetti. After they had been in Georgia for several months, I asked for their salient impressions of the area compared to their Italian homeland. After some disparaging comments about our local food and wine, they both noted amazement at the relative abundance of birds and other wildlife, even on our university campus. Only after I visited Italy could I fully appreciate their sentiments (both on animal life and cultural amenities!). Seeing Italy now, and adjoining countries like Greece, which have a long history of wildlife abuse, it's hard to believe that the region was biotically rich and nearly blanketed by forests 1,000 years ago. These countries today, though picturesque and culturally rich, seem remarkably poor in noncommensal birds and other noticeable wildlife.

On the other side of the Eurasian continent, mainland China may have an even sorrier environmental record. Notably, during the Chinese famine of 1958–1961, Mao Tse-Tung launched a national campaign to annihilate the country's birds and small mammals, which were viewed as crop pests. This genocide was quite "successful," but it only exacerbated human food shortages when insect populations exploded in the absence of control by predators.

New Zealand is another disappointing country in terms of its current avifauna, but for different reasons. Isolated from Australia for eighty million years, the North and South Islands evolved a diversity of endemic birds including many flightless forms: ducks, rails, kiwis, swamp hens, wattled crows, parrots, and even a tiny flightless wren that flourished in the absence of mammalian predators. Especially impressive were about twelve species of moas, some individuals reaching twelve feet in height! Alas, by the seventeenth century, the moas were extirpated by the Maori people and their introduced rats and dogs. Added insults to New Zealand's native biota came from nineteenth-century European settlers, who cleared the land, drained swamps, shot wildlife, and through their "acclimatization societies" purposefully introduced more than 100 animal species from Europe. Exotics now constitute nearly one-half of New Zealand's sixty-five remaining avian species and all of its nonflying mammals (two native bat species still survive).

One need not travel outside U.S. boundaries to see the consequences of faunal turnovers through human actions. The Hawaiian Islands formerly were home to the most spectacular variety of land birds ever known on a remote archipelago, including fifty species of honeycreepers found nowhere else. About 50 percent of Hawaii's native birds were extirpated shortly after Polynesians arrived more than a millennium ago, and one-half of those surviving species then went extinct following European colonization in the early 1800s. Of the few remaining endemic species, roughly 50 percent are now endangered. A visitor to Hawaii today will be greeted instead by a menagerie of "pet-store" birds introduced from around the world. The average tourist doesn't know or care, but most ecologists view this faunal turnover as a biological tragedy of the first order.

A different kind of surprise was in store for me in 1983 when I visited New Delhi. Despite India's crowding, wildlife remained in evidence. Many small birds, for example, were common and tame, suggesting that they view people as little more threatening than cows or camels. Thus, many forms of wildlife apparently can accommodate a strong human presence if not otherwise persecuted. Up to a point, it is not human numbers per se that impacts biodiversity, but rather our social attitudes and actions with regard to the environment.

I still teach ornithology with a conservation bent, and some of my most positive feedback has come from this class. I cherish letters from several former students who, sometimes years later, wrote that the course opened their eyes to nature, giving welcome direction to their careers or enriching their personal lives. I have another reason for continued involvement in ornithology. When I first came to the University of Georgia, bona fide experts taught classes in natural history, ichthyology, mammalogy, and so on. As these senior faculty retired, their replacements often were chosen from a new generation of cell biologists, physiologists, biochemists, and "hard-core" molecular biologists, whose research had greater prospects for attracting significant grant support. Courses on natural organisms were dropped or, as in ornithology, adopted by people like me not specifically trained in these areas.

The net effect was that the scales were tipped too far toward reductionist biology and away from the environmental and natural-history side. Here at Georgia, as at many other universities around the country, it would be entirely possible for a student to obtain a biology degree without ever seeing a living organism in nature as a part of a class experience. Through my own teaching, I take it as a personal responsibility to help rectify that deplorable state of affairs.

10 Corals and Sponges

As our research began cooking and my lecturing improved, graduate students were increasingly attracted to my lab's offbeat kind of research: applying molecular-genetic markers to questions in natural history, behavior, and evolution. One day, Joe Neigel stopped by my office with a proposition. He had been a diving assistant on a coral reef research project in Jamaica, where he had become fascinated by the staghorn coral, *Acropora cervicornis*. Joe now wanted to use genetic markers to examine spatial structure in this species and thereby reveal hidden secrets about its reproductive biology.

As its name implies, each staghorn colony is a branched structure resembling the antlers of a prime buck. Thousands of colonies may cohabit a reef, sometimes packed into "haystacks" of interlocked sprigs. Each colony is like an apartment complex inhabited by hundreds of polyps that have secreted about themselves a protective calcareous matrix, which gives the tenement its distinctive shape and brittleness. As a colony grows, new polyps are budded asexually. Thus, all coral polyps within a given colony are clonemates, ultimately tracing back to one sexually produced larva that settled from the water onto that reef site years earlier.

However, the total number of physically discrete colonies on a reef might be greater than the number of genetically distinct colonies. This is because coral branches sometimes break off from the parent colony, fall to the substrate, and regrow into a new module. Cases of asexual reproduction via colony fragmentation would not necessarily be apparent to a diver but should leave a genetic signature. Joe wanted to employ allozyme markers to score numerous staghorn colonies for genetic identity versus nonidentity and thereby generate spatial maps of particular clones. This exercise might also permit him to deduce the relative frequencies of asexual and sexual reproductive recruitment in the population. Perhaps dozens of neighbor corals had derived from repeated cycles of colony growth and fragmentation over recent decades. At the other extreme, each physically discrete colony might have arisen from a separate sexual propagule.

Joe's overture piqued my curiosity, and his diving tales reminded me of how much I missed the undersea world. My summers in the Bahamas had given me diving experience, but I wasn't formally certified, nor did I own SCUBA gear. So, I soon enrolled in a training course offered by a local dive shop. Most of the lessons took place in a swimming pool, but the final exam entailed open-water dives in the Florida Keys. I felt like a pro among the other students, who were facing the ocean depths for the first time.

My favorite dive was at night. From our forty-foot boat, anchored six miles off Key Largo, we jumped two by two into a black, rolling sea. Our spotlights punctured only a small hole in the inky abyss as we glided to the reef forty feet below. There we were greeted by a fairyland of colors and shapes. Thousands of coral-polyp tentacles danced in our beacons. Gorgonians (soft corals) swayed gently to the ocean's pulse, as did colonies of fan worms protruding from their cylindrical burrows like feather dusters. Nocturnal mollusks, brittle stars, and

spiny lobsters were captured by our roving beams of light, as were solitary and encrusting sponges in reds, greens, yellows, purples, and browns. Fish bumped off drunkenly when disturbed from their slumber. Some made themselves available for closer inspection, such as bright red squirrel fish hovering in coral crevices, and parrotfish, who cloak themselves in pillowy mucous beds of their own secretion. I love night dives.

Back at the lab, Joe and I soon attempted to develop genetic assays for fresh *Acropora* samples, but there were problems. The slimy coral polyps were hard to extract from their calcified skeletons, and they were contaminated by symbiotic algae that live inside the corals' digestive tissues.

Despite our efforts, these technical snags conspired to defeat the molecular-genetic side of our project. But during background reading, we had stumbled upon an interesting literature on histocompatibility-like responses in corals, sponges, and several other invertebrate groups. Many of these organisms possess at least some physiological capacity for the recognition of foreign tissue. For example, sea anemones in tide pools of the Pacific Northwest reproduce both sexually and asexually, and adjacent clonemates on a rock normally interact amicably, whereas neighboring members of different clones fight border wars, using their specialized tentacles with stinging cells. Analogous reports on some marine sponges had demonstrated that artificial tissue grafts from donor to recipient colonies often were rejected, whereas self-grafts were "accepted" as the tissues fused.

In general, such phenomena are reminiscent of immunosurveillance systems, which distinguish self from nonself in vertebrates. For example, artificial skin grafts from one mouse to another normally fail, whereas self-grafts or those among highly inbred individuals typically succeed. These responses are mediated by highly polymorphic histocompatibility genes, which in effect uniquely mark each genetically distinct individual in an outbred population. However, despite intensive molecular searches, invertebrate homologues to these genes had not yet been found. Also, there was ongoing debate about the degree of genetic specificity reflected in the various tissue responses of invertebrate animals.

Two opposing schools of thought were prevalent. According to W. H. Hildemann and colleagues, the graft responses in sponges and corals were exquisitely precise, such that tissue rejection usually meant nonidentity, and tissue acceptance registered clonality. The type of evidence was illustrated by the following field experiment on the reef sponge, *Callyspongia diffusa*. All 739 artificial grafts between widely separated colonies produced rejection responses, whereas those from sponges less than ten meters apart often fused. These neighboring sponges were presumed to be clonemates, but this was not documented by independent genetic evidence.

An artificial graft between two sprigs of staghorn coral.

In opposition to Hildemann's view stood the immunologist A. S. G. Curtis and collaborators, who in their studies of other reef species noted apparent discrepancies in clonal assignments between graft responses and genetic markers. Joe and I were initially impressed by the apparent sophistication of Curtis's laboratory assays and were mindful of design flaws in some of Hildemann's descriptive field protocols. However, we also thought that more refined field experiments could contribute to a resolution of the debate.

Critical support for the self-recognition hypothesis would have to demonstrate that multiple grafts between portions of the same colony invariably fuse; replicate grafts between the same pairs of colonies yield consistent outcomes; each colony can simultaneously accept and reject different third parties, such that the responses don't merely evidence the presence of invariant "acceptor" and "rejecter" colonies in the population; and physiological connection per se is not prerequisite for colony fusion. In addition, a true self-recognition system must exhibit reciprocity, such that the same response is observed regardless of which of two colonies serves as donor or recipient; and transitivity, such that if colony

A fuses with B, and B fuses with C, then A must fuse with C. All of these kinds of field tests seemed eminently doable.

We were soon headed for Discovery Bay, east of Montego Bay on Jamaica's northern coast. This was familiar territory for Joe but new for me. Adjoining the embayment that gave the town its name was a small marine station with several small boats, an air compressor to recharge dive tanks, research aquaria with flow-through seawater, and an apartment complex for visiting scientists. The lab was about a mile from town and overlooked the lovely bay itself and an adjoining barrier reef.

The coral reef that arches along Discovery Bay also parallels much of the north coast of Jamaica about a quarter mile offshore. An inshore back-reef protected by this fringing arc is a sandy area dotted with small coral heads and carpeted by sea grass. Usually six to fifteen feet deep, it offers perfect snorkeling. The fore-reef, on the ocean side, is where the most impressive coral formations occur, before the seafloor slopes away to great depths.

Beginning at the reef summit and proceeding seaward, one first encounters the awesome elkhorn corals, their colonies vaulting skyward, sometimes reaching the ocean's surface. They are close relatives of the staghorn coral and have a similar appearance, except that the polyp-laden branches of elkhorn corals are broader and flattened, actually more like a moose's than an elk's rack. It takes a heavy structure to withstand ocean waves that continually break over these shallow-water stockades.

Interspersed among the elkhorns and also found in deeper waters are "head" corals, which look like decorative seascape boulders. These too are colonial structures, in which clonal polyps occupy the outer surface of calcareous apartment complexes of their own design. Largest are the globular star corals, which can reach the size of a Volkswagen Beetle. Even more intriguing are the brain corals, whose convoluted, channeled surfaces bear eerie resemblance to the hemispherical lobes of the human cerebrum.

Many other corals abound in the fore-reef environment. There are the dreaded fire corals, otherwise inconspicuous, which make themselves known to divers by delivering a searing burn from the slightest touch. There are the many lamellar species of *Agaricia,* terraced in large horizontal plates or standing upright as delicate fronds resembling leaf lettuce. Dense colonies of *Madracis* and *Porites* poke their stubby fingers skyward. Rose corals add their decorative touches to the reef, as do marble-sized *Favia,* tubular *Oculina,* and a host of other smaller calcareous beauties. Altogether, the stony corals leave a wide range of visual impressions, from the subtle delicacy of flower corals to bold phallic pronouncements of the pillared *Dendrogyra.*

Our study species, *Acropora cervicornis,* occupies the midzone of the fore-reef, typically in about twenty to fifty feet of water. The staghorn stands at Discovery Bay were among the most spectacular in the entire Caribbean, forming fields so dense that branches of adjacent colonies often came into natural contact. We soon became adept at scoring these contacts as "fusions" or "rejections." In a fusion response, soft tissues as well as the calcareous matrix have grown together in seamless union, whereas the contact zone in a rejection response resembles a raised contusion, like a soldering joint. Nonetheless, both the fusion and rejection reactions cement the staghorn colonies together in a locked embrace. Indeed, interclasped colonies form a latticework propped above the substrate, up in the sunlight, where the corals' symbiotic algae can do their photosynthetic business. This makes ecological sense, because the last thing a staghorn colony wants is to fall to the seabed, where it likely will die from siltation or the activities of parasitic borers, such as some snails.

In addition to scoring hundreds of natural contacts between coral branches, Joe and I initiated artificial grafts on the reef. The procedure was to break off about a four-inch section of donor staghorn, place it parallel to a branch of the same or another recipient colony, and tie the two branches together using monofilament line attached to a numbered identification tag. Over several weeks, we set up hundreds of grafts in experiments designed to test the clonal-assay hypothesis by addressing the six criteria mentioned earlier. We would return to Jamaica in about eight months to score the matured graft responses.

Joe and I dove twice a day, in the early morning and late afternoon. With our gear piled into a Boston whaler, we motored the mile or two from the lab's dock across the bay and through its inlet to a mooring anchored in fifty feet of water on the fore-reef. In our tiny boat, often dwarfed by ocean swells, we donned the SCUBA gear and somersaulted over the gunwale to be mercifully freed from the weight and jostling above water. My favorite part of the dive was the deliberate, floating descent through clear waters to the reef below. Like spread-eagled skydivers falling through a viscous medium, we gradually approached the stunning seafloor.

Each dive lasted an hour and a half. We spent most of that time hovering in a vertical head-down position, preoccupied with typing grafts, interpreting natural branch contacts, and making notes on our dive slates. We had to position ourselves carefully so as not to bump the fragile coral forest at our fingertips. When scoring artificial grafts from an earlier trip, we also had to search hard for our homemade tags, now encrusted with algae. To any fish in the area, we must have looked like weird, bubble-blowing creatures, coming and going from some other planet.

Two dives a day was just right, and we could not have worked longer because of nitrogen buildup in our blood. Furthermore, we conducted nearly all of the experimentation on the reef itself and thus had little laboratory drudgery upon returning to shore. After cleaning up the dive gear and refilling the tanks, we only had to transfer our field notes, and cut and punch-code plastic tags for the next day's efforts. So, following each late-afternoon dive, we retired to sip daiquiris on our apartment verandah and enjoy the streamertail hummingbirds, likewise sipping liquid refreshment at nearby flowers. In the background, the light of the setting sun danced across the wave-capped sea.

Our working expectation had been that the grafting results would be inconsistent with Hildemann's views and we could then publish the data as strong experimental support of Curtis's contention that marine invertebrates could not possibly have precise enough histocompatibility systems to register clonal differences. However, to our amazement, all of the grafts produced outcomes consistent with expectations of the clonal-recognition hypothesis. The double-blind nature of our experiments made this finding especially impressive. Our tags had been applied to corals whose exact positions could not possibly have been remembered when we scored each graft response underwater a year later. Yet, back on shore, when we cleaned the encrusted tags and looked up their numbers in our log book, each graft outcome proved to be in perfect accord with expectations for a clonal assay.

We were initially dismayed by this result, but our attitude shifted when we remembered our original goals, which had been to identify coral clones and map their spatial arrangements. Although the allozyme assays had failed, we now realized that we might have a natural, in situ bioassay that could accomplish the same tasks. Accordingly, during our second summer in Jamaica, we conducted a more extensive series of artificial grafts designed expressly to map particular coral clones on the reef and describe what might be their spatial patchwork. Through weeks of effort, we set up 1,000 more grafts, planning to return the following summer to score the outcomes.

Alas, our grandiose effort was in vain. About a month after we left Jamaica in the summer of 1980, Hurricane Allen hit Discovery Bay head-on. One of the strongest storms on record, it scoured the bay to a depth of sixty feet and more. The reef was pulverized. No coral survived, and the rubble from the former living museum was soon blanketed by a monoculture of algae. To this day, the reef has not recovered.

Despite this setback, in the end we managed to turn the catastrophe to partial advantage by relocating our experimental efforts to a marine laboratory in St. Croix, U.S. Virgin Islands. Similarly equipped to support diving activities, this lab was also adjacent to a fringing reef that included staghorn colonies. Back in business within a year, we renewed our grafting and eventually succeeded in mapping presumed clones and deducing the coral's population genetic structure.

The clonal diversity of staghorns in St. Croix proved to be far lower than what we had documented in Jamaica. Thus, clonal proliferation via colony fragmentation appeared to have been common in St. Croix, whereas sexual proliferation predominated in Discovery Bay. One plausible explanation is as follows. The fore-reef in Jamaica is narrow, bisected by sand channels that quickly sweep loose sediments to oceanic depths. The St. Croix reef was on a broad shelf, heavily silted by comparison. Tiny coral larvae, the products of sexual reproduction, require a hard surface to settle. Because attachment sites for colony initiation were less restricted in Discovery Bay, sexual recruitment was probably more common there. Joe ran computer simulations that bolstered this notion: suspected rates of sexual versus asexual recruitment into the simulated populations closely predicted our empirical estimates of clonal diversity at the two sites.

Hurricane Allen also provided an important ecological lesson. It drove home the point that periodic environmental disturbances can have profound impact on a reef's makeup. A perusal of the literature suggested that singular disasters such as a hurricane, cold snap, or disease epidemic deal a devastating blow to any particular Caribbean reef site about once every century, on average. Indeed, the St. Croix reef was damaged by a strong hurricane just a few years after we completed our work there. In another example, the long-spined sea urchin was abundant on Caribbean reefs during our studies, but a waterborne disease later caused nearly 100 percent mortality in the species.

Such examples led us to appreciate that the kinds of ecological and demographic factors monitored by natural historians at "normal" times often paint an incomplete if not misleading picture of what can influence the longer-term genetic architecture of a reef. For example, in interpreting population genetic structure in the staghorns, customary considerations would include annual rates of colony fragmentation and regrowth, larval settlement, and routine sources of colony mortality, all plugged into equilibrium models predicting clonal diversity in the species. Hurricane Allen taught us that unique events also must be factored into the equations. During my career, I have become increasingly convinced that most natural populations (in the marine realm and elsewhere) fluctuate dramatically in size, if only periodically, and that equilibrium models based solely on normal year-to-year parameter values often are inadequate to capture the overriding genetic impacts of idiosyncratic historical events in populations. Indeed, one of the finest aspects of genetic analysis is that it often permits biologists to deduce historical impacts that would not necessarily be evident to a contemporary observer.

St. Croix and Discovery Bay also differed in ways that affected our diving pleasure. For example, Jamaican waters were nearly devoid of sizable fish. For many years, islanders had hooked, netted, speared, dynamited, poisoned, and trapped reef fish for food. Meat on the island was scarce, as we learned on our first visit when we had to queue each Thursday for our weekly allotment of one scrawny chicken from the village grocery store. Like the locals, we mostly ate rice, fruits, vegetables, bread, canned sardines, and Jamaican hot sauce. By the late 1970s, the islanders had decreased the mesh sizes of their fish traps such that even puny fish were taken for chowders. Rarely did we see any reef fish longer than about two inches.

St. Croix, by contrast, was fish paradise. Joe and I were accompanied on our daily dives by four-foot barracudas, herds of goatfish, angelfish, triggerfish, wrasses, butterflyfish, blue chromis, sergeant majors, grunts, surgeonfishes, and many others. Like us, trumpetfish often hovered vertically, their intent being to sneak up on smaller fish and nab their prey with lightning quick thrusts. Also like us, parrotfish had a coral fetish, but their interests lay in munching the hard colonies with beaklike incisors. Three-inch-long dusky damselfish also were coral devotees. Their agrarian lifestyle involves defending carefully cropped algal gardens that grow on the skeletons of deceased staghorn corals. We received many a nip from these feisty little assailants, whose diminutive size belies their combativeness. Thank goodness barracudas don't have the same aggressive temperament!

It was also somewhat harder for us to reach the St. Croix dive site. Because of the reef's configuration, we had to anchor our little boat on the back-reef and swim to the fore-reef through a shallow cut at the reef crest. On rough days, this passage was dangerous, and we had to carefully time our dashes between crashing waves. One day my weight belt snagged on a coral head, and I didn't make it across before the next roller arrived. For several agonizing seconds, I was dashed like an inconsequential toy against coral heads and spiny urchins. Miraculously, I emerged unscathed, my tank and buoyancy vest having taken all the serious blows and stabs.

On our dives, Joe and I noticed several species of ropy sponges with growth forms generally reminiscent of the staghorn corals. Our favorite went by the imposing name of *Iotrochota birotulata,* a velvety green sponge scattered among coral heads. This species grew in sprawling aggregations of foot-long branches that sometimes curved back on themselves and fused. Could this species also display acceptance and rejection reactions that would provide clone-specific bioassays in artificial grafts? We came to believe that the answer is yes.

Our underwater grafting procedures were much like those we had employed for the corals. Within about two weeks, grafted branches either fused seamlessly or else had rejected one another. Collectively, the various response patterns in this sponge were entirely consistent with expectations of the self-recognition hypothesis. Accordingly, we again used the grafting results to map the spatial arrangements of clones. Each clone proved to be confined to a single patch less than three meters in diameter. There, the sponge sometimes existed as multiple colonies that formerly must have been interconnected but apparently then became sundered by wave action or the death of intervening tissue.

We also attempted grafting experiments with several other sponges and corals. However, for technical reasons not all species could be grafted, and those that did sometimes yielded unscorable outcomes. For example, artificial grafts in one sponge species appeared to fuse initially but partially reject later on. Such complications, as well as probable differences among invertebrate species in the degree of specificity in the grafting response, probably account for why the Hildemann-Curtis debate had remained so difficult to resolve.

Joe and I weren't entirely preoccupied with research. We took breaks to visit Montego Bay, the "cockpit country" of Jamaica's interior, and other tourist sites. We enjoyed the relaxed island lifestyle and the dreadlocked Rastafarians with their "hey mon" greeting and large, smoking reefers.

One such lion-maned, red-eyed fellow scared the heck out of us, however. We had made arrangements with a tour company for a weekend visit to Jamaica's highest peak, in the Blue Mountains near Kingston. This required the services of a four-wheel-drive jeep. At the mountain's base, we were greeted by a chauffeur with a giant reefer in one hand and a beer in the other. To constant assurances of "no problem," he drove us up toward base camp. The "road," sandwiched between a vertical rock wall on one side and a sheer drop-off on the other, was so narrow that I swear the right tires partly overhung the precipice on several occasions. The driver must have been smoking some really good ganja because he got us there alive and was nonchalant about the miracle. Only in retrospect was the adventure worth the anguish, because I did add several species of high-altitude Jamaican endemics to my avian life list.

Joan accompanied us on some of these trips to the Caribbean. She loves to collect shells and enthused me about the hobby. Over the years, we have collected and identified most of the 1,000 mollusk species inhabiting the western Atlantic and the Caribbean, and, in a sort of reversion to my childhood, I have labeled and housed them in banks of organizer drawers in our home. In Jamaica, Joan and I spent many a pleasant hour wandering the shorelines in search of gastropod and pelecypod specialties. I collected a few live specimens on dives, also, although mollusks seldom are observed underwater because they mostly lie buried or hidden in the daylight hours. My favorite species from

Discovery Bay was the spiny oyster, whose shells of pink, cream, and brown are festooned with long, delicate fronds. These oysters were common at sixty feet on the pilings and canyon walls in the murky waters beneath the loading dock for bauxite ore boats, which squeezed into Discovery Bay.

Joe and I also dove recreationally on occasion. On clear nights of the full moon, for example, we sometimes returned to the fore-reef to lie motionless in white sand channels at fifty feet. The greenish moonlight, filtering down through the staghorn forest, reflected off our bubbles and cast ghostly shadows on the coral heads. This mystical setting was so relaxing that it took some concentration not to fall asleep in this gentle oceanic womb.

On another diving adventure, Joe and I followed a nearby canyon wall to a depth of 170 feet. On our left side was the cliff face, cloaked in corals, gorgonians, and sponges. Especially eye-catching was a species of orange sponge that gave the appearance of having dripped from one rock ledge to the next because it formed nicely aligned biotic stalactites and stalagmites. Also hugging the wall and its overhanging ledges were lovely royal grammas, brilliant purple-and-yellow fish who, with equal facility, swim right side up or upside down in their gravity-free world. On our right side was the open ocean, from which pelagic fish strangely materialized and then disappeared again in the inky blue beyond. Below us was black nothingness. Looking down, my stomach came into my throat as it does when I peer over the edge of a tall building. Looking up, I could barely trace the long trains of ascending bubbles as they hurried toward liberation in the distant atmosphere. In the rapidly fading light of the depths, my mind became a drunken blur from otherworldly sensations and the gentle touch of nitrogen narcosis.

Like a jolt from a surreal dream, Joe grabbed me by the arm and motioned that it was time to ascend. Likewise, after each marvelous trip to the Caribbean, it was a shock to return to the world of academic life in Georgia.

11 Energized by Mitochondria

By the late 1970s, scientific publications were flowing steadily from my laboratory. They described genetic findings on the ecology, behavior, or evolution of natural populations in a wide variety of species, typically as revealed by allozyme markers. This field of molecular natural history was embryonic, and I was one of its few devoted practitioners. Yet, my group had not achieved any breakthroughs that might be deemed extraordinary. That would soon change.

In 1977 I gave a departmental seminar describing comparative genetic patterns in vertebrate animals. I concluded my talk by echoing a popular sentiment of the time: regulatory genes (rather than those encoding cellular "housekeeping" proteins) should be examined next, because changes in gene regulation might be at the heart of adaptive evolution. One person in the audience asked if restriction enzymes might be employed to study stretches of DNA involved in the regulation of gene expression. I had never heard of restriction enzymes, but this casual question planted a pivotal seed in my mind, which soon would germinate and eventually grow into an altogether different kind of genetic revolution—a revolution not in the study of gene regulation, but rather in new ways to study natural history.

Restriction enzymes were to be an important part of the foundation of that revolution. Directed to published studies, I soon read that these enzymes are like refined molecular scissors that precisely snip DNA from any species at specifiable points. Bacteria produce restriction enzymes and in nature use them to defend themselves against invading viruses by slicing up the infective viral particles. In science laboratories, these biochemical scissors would soon be harnessed by geneticists for manipulating plant and animal DNA in a host of applications that today are an important basis of the biotechnology industry and "genetic engineering."

I had no desire to engineer genes, but I did begin to wonder if restriction enzymes might somehow be used to reveal DNA-level markers at sites of gene regulation. If so, I could move from studying enzymes (the products of DNA) to DNA itself. I soon visited with several faculty in the microbiology and biochemistry departments at Georgia to identify a collaborator from whom I might learn the appropriate lab techniques. My inquiries led nowhere, with one exception. Microbiologist Bob Lansman welcomed me to his laboratory but also apologized that he had little experience with regulatory DNA (most of which is housed in the cell nucleus). Instead, his career had been devoted to the biochemistry and cellular biology of mitochondrial DNA (mtDNA), housed in the cell cytoplasm. Again, I had barely heard of mtDNA, but this little molecule would soon prove to be another important part of the foundation for my first true revolutionary breakthroughs in science.

Directed again to published studies, I soon read how mitochondria, a cell's miniature power-generating plants, produce and store the energy that each cell uses to drive its biochemical operations. I also learned that within each mitochondrion are multiple mtDNA molecules, each a tiny circle of about 16,000 pairs of nucleotides (the basic building blocks of all forms of DNA). Astonishingly, all or almost all mitochondria are thought to be maternally transmitted across animal generations, such that each offspring inherited its mother's (and not its father's) mtDNA genotype. This is because when an egg and a sperm unite to begin a new individual, most of the cytoplasm, and thus mitochondria, in the fertilized egg derive from the female parent.

Another curious fact is that mtDNA bears remarkable structural and functional similarities to bacteria. Indeed, a consensus scientific view is that mitochondria trace back in evolutionary time to bacterial ancestors that first entered into "endosymbiotic" relationships with host cells more than a billion years ago. Symbiosis describes any case in which dissimilar organisms live in close association, often to mutual benefit. The prefix "endo," from the Greek for "within," implies an intimate alliance. This ancient microbial wedding produced a lasting marriage. Literally, the cells of higher organisms (ourselves included) are an integrated mosaic of nuclear and mitochondrial genes that long ago were housed in independent presymbiotic microbes. Yet, we could no more live without our internal bacteria-like associates than they could live without us. Indeed, "we" and "they" are one.

Despite the fascinating evolutionary properties of mtDNA, I initially viewed it merely as a learning vehicle for restriction-enzyme assays. Soon, Bob and I were purifying and digesting ("restricting") mtDNAs from mice and other rodents and then separating the resulting digestion products through agarose gels (with electrophoretic methods quite analogous to those I had used for years to separate proteins).

This effort, which began merely as a training exercise in restriction-enzyme methods, soon opened a world of novel biological questions that could not be ignored. Why did each individual display only a few mtDNA bands on a gel? I thought it must be because most of the trillions of mtDNA molecules in an individual were identical in genotype, such that they were cut at precisely the same molecular sites by a given restriction enzyme. Otherwise, if the mtDNA molecules within an individual differed genetically from one another, they would be cut at many different restriction sites, resulting in a smear of mtDNA fragments of varying size on a gel. Why then did different mice in a population often display distinct mtDNA fragment patterns? It must have been because mutations arise and sometimes precipitate a rapid turnover in the population of mtDNA molecules that a mother transmits to her progeny.

The fact that large populations of mtDNA molecules inhabit each cell also meant that a whole new level of genetic analysis was begging for study. One of my graduate students, Bob Chapman, soon began mathematically modeling alternative scenarios concerning the population genetics of mtDNA molecules within somatic and germ lines. Using novel data and theory, we thus began to explore the unknown intracellular dynamics of mtDNA populations. Other intriguing questions arose at the level of animal populations. What was implied by the fact that different mtDNA genotypes within and among related species appeared to be genealogically linked?

The mitochondrial revolution, for us, began quite suddenly when we came to appreciate that mtDNA molecules record the matrilineal histories of successive mutations. These mutations reveal phylogenetic (genealogical) connections among what in effect are female family names. An excellent analogy is provided by male surnames in some human societies. Just as sons and daughters "inherit" their father's surname, which only sons can pass on, so too do males and females normally inherit their mother's mtDNA genotype, which only daughters can transmit to their offspring.

Furthermore, much as "mutations" sometimes arise in human surnames (my own family name was a nineteenth-century misspelling of Avis), small mutations frequently occur in mtDNA genotypes. Thus, mtDNA molecules record matrilineal histories of descent much as surnames trace patrilines. However, two salient differences give mtDNA a compelling advantage in historical reconstructions: surnames were invented only within the past few centuries, but mtDNA's matrilineal registry goes back a billion years in evolutionary time; and only humans have surnames, but nature has provided a mtDNA family name for essentially every multicellular animal. We had, in essence, found a new key to unlock many of evolution's mysteries.

Tantalizing questions continued to emerge from our initial genetic observations on mtDNA. Why did members of each sexually reproducing species tend to group together in terms of matrilineal history when the conventional evolutionary bonds of mating and genetic recombination seemed at face value not to apply to these asexually transmitted mtDNA genes? What might be the ramifications of a novel view treating each distinct mtDNA genotype as a matrilineal "clone," and each individual (rather than a population) as a basic unit of genealogical analysis? Indeed, what meaning did "phylogeny" have at the intraspecific level?

Years would pass before fully satisfying answers to these and many related questions would be forthcoming, but I began to see that we had stumbled upon an untapped bounty of unorthodox conceptual as well as empirical perspectives in evolutionary biology. I soon forgot about my original but less tractable goal of studying regulatory genes and shifted all available resources, instead, to these exciting matrilineal markers.

Our first mtDNA population-level studies involved *Peromyscus* mice and *Ge-*

Photographing a pocket gopher in southern Georgia, 1979. The animal is barely visible at the top of the mound.

omys pocket gophers, creatures chosen mostly for their local availability. One species of interest was the beach mouse, *Peromyscus polionotus,* a fairly common inhabitant of the southeastern coastal plain, where it replaces its sister species, *P. maniculatus,* which inhabits the remainder of the continent. Beach mice are specialists on sandy soils. They dig deep burrows, typically occupied by an adult couple or by a mother and her litter of two to five pups.

The tubular burrow, two inches in diameter, usually is dug into an exposed slope, such as along a road cut. Slanting downward for about four feet, it then opens into a spherical nest cavity eight inches in diameter. Leading back from the nest are one or two escape tubes extending upward to within an inch of the soil surface. When mice are pursued in their burrows by a snake or weasel, they flee via these escape routes, eventually "popping the cork" to emerge above ground at distances often ten feet or more from the main burrow entrance.

If I do say so myself, I am a skilled connoisseur of the beach-mouse collecting procedure. I cruise some country road looking for a sandy mound marking a burrow entrance. Even at 55 mph, with practice it is possible to distinguish mouse earthworks from those of pocket gophers, tortoises, and fire ants. Once a burrow is located, I excavate it slowly and carefully by shovel, using successive vertical slices through the tunnel, about one inch apart. If the mice are home, a firm sandy plug is encountered less than a foot inside the entrance. This plug, about three inches long, differs in color and texture from the surrounding soil. Without a plug, there is no point in continued digging, because the nest is unoccupied.

Confidently digging past a plug, inch by inch, I take care not to lose the meandering tunnel while at the same time shoveling away the embankment so that it doesn't collapse on my trench. Soon, the nest cavity is encountered. A quick sniff of its grass lining confirms that mice were recently present and now are on their way up the escape tubes. I locate a tube and follow it by continued digging. When I get to within about two feet of the soil surface, the mice pop out nearby, and I pounce on them with headlong dives. The squeaking rodents, one or two in each hand, are then safely placed in cages.

The southeastern pocket gopher *(Geomys pinetis)* offers another collecting challenge. These creatures, resembling a short loaf of French bread, live in burrows of an entirely different design. From above, each tunnel network is recognized by a cluster of sandy mounds lacking surface openings. Choosing one such hillock, I excavate a cylindrical hole about three feet deep and four feet in diameter. This cavern typically intersects the gopher's underground tunnels at two or more points, into which I stuff spring-operated live traps made from PVC pipe. Within hours, an unsuspecting gopher may amble into one of these de-

vices. Removing the animal takes care, because its sharp incisors can deliver a nasty bite.

Professor Joshua Laerm, late director of the university's Natural History Museum, taught me proper gopher-collecting etiquette on several field trips to South Georgia. Josh was a paleontologist, mammalogist, philosopher, environmental spokesman, redneck, and former member of a motorcycle gang. Tough as nails and always politically incorrect, this irreverent closet intellectual also had a deep affection for nature, and we became great friends. We loved getting all mean and sweaty while drinking beer and digging gophers in some ugly pasture under the blazing Georgia sun, arguing politics and damning what human overpopulation was doing to our planet. Josh passed away a few years ago while in his prime, before fulfilling his dream of building a premiere university museum devoted to environmental issues. His untimely death reminded me of the fragility of life and in no small part convinced me to write this autobiography while still fairly young.

Josh and I collected eighty-seven pocket gophers from sites in South Georgia, Florida, and Alabama. In Bob Lansman's lab, mtDNA molecules were purified from these animals and digested with several restriction enzymes that

Bob's graduate students had isolated. In those days, few restriction enzymes were commercially available, so researchers had to purify their own from large bacterial vats, sometimes trading the excess among themselves for other enzymes or laboratory supplies. Similarly, twenty years earlier it had been customary for graduate students in biochemistry departments to isolate various ingredients used in their protein stains. Later, these biochemicals became available from commercial outlets as a matter of course, as did restriction enzymes. This is a recurring theme in science, wherein a breakthrough technique of one era becomes a mundane support technology of the next.

Tremendous population genetic variation was evident in the mtDNA profiles of our mice and pocket gophers. This surprised everyone greatly, because at that time molecular biologists generally assumed that mtDNA would prove to be exceptionally conserved and slowly evolving, given the molecule's genetic "economy." MtDNA is small and tightly packed with genes coding for proteins and other molecules centrally involved in metabolic transactions critical to cell function. A central tenet of molecular evolution is that functionally important genes evolve slowly and display less structural variation than those less subject to functional constraint. For some reason, mtDNA molecules seemed to violate this paradigm.

Shortly thereafter, in 1979, an influential article was published by a young postdoc, Wes Brown, working in Allan Wilson's lab at Berkeley. This paper announced a rapid pace for mtDNA sequence evolution in mammals, an observation later confirmed and extended to many additional animal groups. Why mtDNA evolved so rapidly was a puzzle, but several hypotheses soon were raised. Perhaps mtDNA molecules experience a high mutation rate, due for example to fast replicative turnover within cell lineages, inefficient mechanisms of mtDNA repair, or exposure to intracelluar mutagenic agents. Alternatively, perhaps the molecule is not unduly constrained in function after all (it does carry few genes, and essentially none that are involved in its own replication or expression). Today, all of these explanations appear to have some validity.

Wes Brown (apart from ourselves) was the other principal pioneer of mtDNA studies at the microevolutionary level, and his story also illustrates serendipity's role in science. Wes had been a graduate student at Cal Tech, in a lab studying mtDNA transcription and physical chemistry. In 1971 he visited an exhibition of M. C. Escher paintings at the Los Angeles County Museum, where he happened to meet John Wright, curator of the museum's herpetology department. Wright was a leading expert on *Cnemidophorus* lizards, many of which reproduce by the bizarre means of parthenogenesis.

Each parthenogenetic lizard taxon consists solely of females who propagate themselves clonally by laying unfertilized eggs. Each unisexual species arose via one or more hybridization events between a pair of normal sexual species. Brown quickly appreciated that the maternally inherited mtDNA molecule could be an ideal phylogenetic marker to identify the female parental species in each such cross and in general to reveal the genealogical histories of the unisexual clones. Over the next twenty years, Wes and his collaborators examined evolutionary genetic patterns of the *Cnemidophorus* lizards in a wonderful series of studies initially sparked by that chance encounter in a county museum.

In our own lab, genetic analyses of the pocket gophers by various restriction enzymes revealed many different mtDNA genotypes, some genealogically closer than others. For example, certain pairs of genotypes differed by only one assayed restriction site, whereas others differed by a dozen or more such alterations. Pretending, literally, that each distinct genotype was a different female family name, I summarized the data into letter codes.

Only a precious few times in one's scientific career does a true revelation come along. I'll never forget the afternoon when I first sat down to contemplate these coded data. Using scratch paper, I started to draw lines connecting related genotypes. As more mtDNA family names were added, the nexus gradually extended outward until all genotypes were linked. This network parsimoniously described the most likely set of evolutionary pathways along which the mutational differences had accumulated. My scribbled drawing looked ugly, but its underlying beauty quickly materialized when I superimposed the network over a map of the geographic sources of the gopher collections. Eureka!

Each female family name proved to be confined to a particular part of the map, and closely related family names either overlapped in geographic range or were spatial neighbors. Furthermore, a deep genealogical gap cleanly distinguished all gophers collected in the eastern half of the species range from those to the west. In short, the mtDNA genealogy had a striking spatial orientation. My little hand-drawn network superimposed on geography had captured, for the first time in any species, a simple yet elegant picture of intraspecific genealogy.

The public unveiling of these findings took place in the summer of 1978 at the International Congress of Genetics held in Moscow, USSR. I had agreed to deliver a plenary talk on protein variation and speciation in fish, earlier work for which I was better known. However, as the conference neared, I was preoccupied by our genealogical discoveries on mtDNA. As I left for the Atlanta airport, I threw several relevant slides into my briefcase. On the plane, and in the Russia Hotel on Red Square across from the Kremlin, unable to contain my enthusiasm, I hurriedly wrote a lecture on this new topic. When my turn came to speak before a packed auditorium, I announced a title change and spoke on our startling molecular work revealing the maternal ancestries within species.

Everyone immediately grasped the revolutionary significance of this approach.

Soon thereafter, our article, "Mitochondrial DNA clones and matriarchal phylogeny within and among geographic populations of the pocket gopher, *Geomys pinetis,*" appeared in print. Together with our similar analysis of beach mice, also published in 1979, this work laid the foundation for a field I termed "phylogeography," which was to grow explosively during the 1980s and 1990s, much as had the allozyme revolution a decade earlier.

Of course, no one cared particularly about the specific findings in terms of the rodents themselves. As with most scientific breakthroughs, it was the documentation of a powerful new conceptual and empirical approach that captured the imagination. There were several innovative aspects to these studies. For example, for the first time individual animals could be treated as basic units of phylogenetic analysis. In contrast to most prior population genetic work based on allozymes, where genetic complications of diploidy, segregation, and sexual recombination necessitated that groups rather than individuals be the focus of attention, the maternal histories of individual animals could be recovered in the mtDNA assays. Thus, by eliminating the need to artificially circumscribe each "population" a priori (by geographic or morphological criteria, for example), an unwanted element of circularity could now be removed from genetic studies of populations in nature.

Our papers also introduced the unorthodox notion that phylogenetic terms and concepts could be applied to evolutionary issues within a species. Before that time (and continuing today in some circles), an entrenched notion was that phylogeny has no meaning at intraspecific levels because conspecific lineages are connected by sexual ties, rather than hierarchically branched. Traditionally, speciation had been viewed as a sharp boundary demarcating the field of population genetics on one side from phylogenetic biology on the other, a perception that has seriously hindered attempts to connect microevolutionary and macroevolutionary thought. An oft-voiced warning was that speciation is a line of death below which phylogeneticists dare not tread. However, due to mtDNA's maternal (and thus asexual) heredity, matrilineal genealogies could now be seen as treelike components of an animal pedigree (a species' true microphylogeny). Thus, mtDNA data gathered within (and among) related species could be interpreted using technical and conceptual approaches traditionally reserved for higher-level phylogenetic biology.

There always had been a profound irony in the conventional state of affairs, which promoted an estrangement between microevolutionary and macroevolutionary studies. The mtDNA data provided a springboard for the explicit realization that these formerly disengaged disciplines must connect, as they inevitably deal with biological phenomena along a temporal continuum of genetic transmission. All limbs and branches in any larger tree of life must have had a substructure consisting of ever smaller branchlets and hereditary twigs that in principle consist ultimately of the generation-to-generation pedigrees of genetic transmission through each species.

This point can hardly be overstated, because throughout most of the twentieth century, population genetics and phylogenetic biology had spoken different languages and gone their separate ways. Typically, a population geneticist might have had a strong background in mathematics or ecology and would view the evolutionary process as changes in allelic or genotypic frequencies under the influence of mutation, gene flow, random drift, natural selection, and the mating system. To a traditionally trained population geneticist, phylogenetic terms such as clades, outgroups, and sister taxa, as well as the methods of phylogenetic inference, would have been quite foreign. Conversely, an old-school phylogenetic biologist would have had extensive training in taxonomy and systematics (perhaps via museum curation), but he or she probably would have little familiarity with or interest in population genetic models or microevolutionary processes.

Nonetheless, as had been noted by the famous paleontologist George Gaylord Simpson in 1945, "The stream of heredity makes phylogeny; in a sense it is phylogeny. Complete genetic analysis would provide the most priceless data for the mapping of this stream." These hereditary streams originate as rivulets in the transmission routes along which particular pieces of DNA have trickled through an animal or plant pedigree, from parents to offspring, generation after generation. Our (and Wes Brown's) introduction of mtDNA approaches to population biology opened the door for scientists for the first time to resolve one such transmission route—the matrilineal pathway—within any species.

My laboratory and others soon began publishing extensively on the novel concept of gene trees as distinct from, but also as constituents of, population or species trees. No longer content to deny any relevance for phylogenetic concepts in microevolution, we saw a new and positive challenge before the field: how to properly interpret the status of gene trees as meaningful components of extended animal or plant phylogenies. Each piece of DNA in a cell's nucleus or cytoplasm truly has a unique phylogenetic history, a recent part of which has been its genealogical transmission route through an intraspecific pedigree.

A related challenge was to examine how population demographic factors (such as means and variances in offspring numbers, fluctuations in population size through time, and other historically dynamic parameters) might relate to the structure and temporal depths of gene trees. With such theoretical relationships worked out, it should then be possible to reverse the train of logic and use empirical gene trees to reconstruct the otherwise unknown demographic histories of populations in nature. In the ensuing decades, such lines of inquiry were to grow into a discipline now termed coalescent theory, whose embryonic roots were planted and nourished in the rich empirical soil of mtDNA analysis. The word "coalescent" refers to the fact that if one looks backward far enough in time, all lineages in extant organisms must trace genealogically (i.e., coalesce) to common ancestors.

Another novel but realistic perspective prompted by mtDNA analyses involved a movement away from traditional equilibrium theory. Beginning early in the century, a sophisticated mathematical body of population genetics had arisen around models of population structure designed more for mathematical tractability than for agreement with nature. For example, a popular "island model" assumes that a species is subdivided into equal-sized populations, all of which exchange genes with equal probability. Few if any species in the wild closely approximate this idealized configuration.

Under such models, traditional theory usually addressed "equilibrium" expectations, such as the population genetic structure anticipated from a long-term balance between the diversifying effect of random genetic drift and the homogenizing influence of gene flow. However, real populations come in all shapes and sizes, experience eccentric gene-flow regimes in time and space, and in general have idiosyncratic histories of demography and distribution. Indeed, evolutionary biology is quintessentially a historical rather than an equilibrium science. Furthermore, it is often of interest to learn the detailed sagas of particular species or populations. Mitochondrial DNA, by virtue of its genealogical clarity, permitted unprecedented access to such historical information over recent evolutionary time.

Thus, empirical work on mtDNA and the rise of coalescent theory have revolutionized microevolutionary studies. Most importantly, they have built empirical and conceptual bridges between the formerly disengaged fields of population genetics and phylogenetic biology. In my opinion, genealogical perspectives on intraspecific genetic variation have gone a long way toward unifying micro- and macro-evolutionary perspectives into a more intelligible and realistic whole.

Quite apart from its genealogical applications, mtDNA has prompted novel insights in cellular biology and medicine. For example, several human genetic disorders, sometimes associated with senescence or aging, have been mapped to malfunctioning mitochondrial genes. As a sidelight, I have also maintained an interest in such matters and in the broader evolutionary ramifications of mtDNA's uniparental inheritance. For example, one paper that I published in 1993 was a novel attempt to integrate the subjects of senescence, sexual reproduction, and the evolution of cellular mechanisms for DNA repair. As detailed in my article, these seemingly disparate topics are in fact thoroughly intertwined.

This paper inaugurated a special "Perspective" section, which continues to this day in the journal *Evolution*.

My intellectual curiosity, not to mention scientific career, was greatly energized by the novel outlooks first prompted by studies of mtDNA. For the next twenty-five years, my students and I were to explore this exciting gold mine and excavate many of its scientific nuggets.

Part Three

A Revolution Applied

I once read that fewer than 10 percent of Ph.D. recipients in biology go on to conduct scientific research that departs substantially from the specific theme or approach of their doctoral dissertation. In my career, by contrast, I've had the good fortune to stumble into a succession of divergent research areas, united only by being within the catholic area of genetics, natural history, and evolution. Two major factors are responsible for the topical and taxonomic breadth of my laboratory's research program.

First, owing to a succession of technological breakthroughs, genetics has been an incredibly dynamic field in recent decades. As each new laboratory technique was invented and deployed, it literally drove studies in ecology and evolution by offering new windows on a host of research problems. First came the protein-electrophoretic revolution of the 1960s, followed a decade later by restriction-enzyme analyses as applied primarily to mtDNA. In the late 1980s came various "DNA fingerprinting" methods, along with technical refinements for direct assay of DNA sequences themselves (long considered by many researchers the Holy Grail of genetic analysis). In my laboratory, we have implemented each of these techniques and others in our exploration of genetic patterns and processes in nature. Any research program active in ecological-genetic or molecular-evolutionary research over the past thirty years must continually have incorporated new technologies.

Second, the essence of inquiry in evolutionary genetics often demands a wide view of the biological world. Genes and their protein products are essentially universal to life, and not to capitalize upon the comparative aspects of this historical as well as functional information would be to fail to read nature's genetic diaries for all they are worth. In 1994 I published my first textbook, *Molecular Markers, Natural History, and Evolution,* and its success reflected in part the book's emphasis on comparative aspects of molecular evolution, which were eye-opening to many ecologists, behaviorists, field naturalists, and other biologists, to whom the work was directed.

The following chapters will offer personal background and anecdotes from my major research endeavors over the years, as recounted in a topical rather than strictly chronological fashion.

12 Phylogeography

Early reports on mtDNA variation from my laboratory and others focused primarily on mammals. These studies demonstrated that many mammalian populations are structured genealogically at scales ranging from closely related families within a forest or meadow, to regional populations separated for millions of years. As described in the preceding chapter, such genealogical treatments afforded many novel perspectives in mammalian genetics and natural history. Would high levels of molecular variation and asexual inheritance characterize the mtDNA of other animals as well? If so, might phylogeographic patterns likewise illuminate these species' natural histories and their recent evolutionary pasts?

Biff Bermingham was the next graduate student to join my lab. Fish were to provide his cast of characters, and the southeastern United States was the evolutionary theater whose phylogeographic plays he would critique. This meant that populations of each of several fish species would be sampled from a dozen or so major river drainages traversing the Carolinas, Georgia, or eastern Florida on their way to the Atlantic Ocean, and those entering the Gulf of Mexico after negotiating coastal states from Alabama to Texas.

Considerable field effort was entailed, often with assistance from fish-and-game personnel. Together we collected fish from southeastern rivers, anxious to explore molecular phylogeographic patterns in any species that turned up consistently. Three such species were the warmouth, spotted sunfish, and redear sunfish, all close evolutionary relatives of my old acquaintance the bluegill. Another fish captured routinely was the bowfin, a living representative of an otherwise extinct taxonomic group that flourished more than 100 million years ago, before the adaptive radiation of "higher" bony fishes. This "living fossil" does look ancient, with a large, bony head, conspicuous scales, an extended dorsal fin, and an asymmetrically lobed tail.

Rivers of the Southeast come in many sizes. Small streams confined to the sandy coastal plain typically run clear, albeit stained a tea brown by decayed vegetation. By contrast, rivers traversing the muddy Piedmont are an opaque red-brown, heavily laden with clay washed from the land. It was not always thus. When the naturalist William Bartram traveled the region in the late 1700s, Piedmont rivers ran clear, even after heavy rains. The unbroken forest that stretched from the Atlantic Ocean to the western prairies provided a canopy and root system that protected against soil erosion. The crystalline waters described by Bartram, teeming with fish, were degraded to today's murky watercourses only after the forests were cleared and the land converted to agriculture following European colonization.

In large rivers, our most effective fishing method was electroshocking. From a small boat housing a noisy power generator, an electrode dangles into the water. To complete the circuit, the operator extends a metal pole into the water and engages the "dead man's" switch. The procedure isn't as gruesome as it sounds. The local electrical field merely stuns nearby fish, which then are dipped up with long-handled nets. Most of the fish recover quickly in holding tanks and seem none the worse for the tingly experience. The same could be said for us dip-netters, who, in the excitement, sometimes got jolted when we forgot to don rubber gloves before lifting a fish from the water.

I mentioned earlier that each graduate student can present a unique challenge for a major advisor, and Biff was no exception. Self-confident, ambitious, headstrong, and boundlessly energetic, Biff got me into several predicaments that I could view with good humor only years later. Biff is now an active senior researcher at the Smithsonian Tropical Research Institute in Panama, and I hope he will forgive me for relating two stories that illustrate his graduate-student style.

On one field trip to Alabama, Biff apparently failed to secure the boat trailer to the ball-hitch of our departmental van. When the trailer's tongue jolted loose on the expressway at 60 mph, the boat and trailer fishtailed wildly, held to the van only by thin chains. The mishap, recounted to me in classic Biff style upon his return, emphasized the positive. In his version, a state trooper who had witnessed the event heartily congratulated him on his consummate skill in getting the trailer under control without killing anyone. Still, I was the one who had to explain to the department chairman why our vehicle now had a huge dent in its rear door.

A second embarrassing incident came to my attention via a midnight telephone call from campus police. A senior faculty member, working late, had complained of a noxious smell in the six-story biology building. Upon investigation, the police discovered the source: a tub full of rotten fish on the roof. Informed that I was the only resident ichthyologist, the police demanded an explanation, but I had no clue. Only after driving to school and discovering that the fish were bowfin did I put two and two together. It turns out that Biff had wanted to preserve bowfin skeletons for the fun of it, so he left fish on the roof to decay naturally. Unfortunately, he had placed the putrid tub of fish gruel right next to the building's air intake. Not to worry—within a few days, the smell gradually dissipated, and everyone quit complaining.

Biff conducted the mtDNA assays on the bowfin, and to our delight found both molecular and population-level features strongly reminiscent of those in small mammals. Each specimen displayed one mtDNA genotype, yet individuals even within local populations often differed from one another by specifiable mutations. The data also revealed a matrilineal phylogeny that mapped beautifully onto the spatial arrangement of rivers.

In particular, a major genealogical subdivision distinguished all bowfin samples taken along the Atlantic Coast and peninsular Florida from those in most of the rivers entering the Gulf of Mexico. Incredibly, the spotted sunfish, warmouth, and redear sunfish showed phylogeographic patterns quite like those of the bowfin, and prominent genealogical subdivisions in each case distinguished similar sets of regional populations. Within each species, the Atlantic coastal plain and peninsular Florida typically housed one set of related genotypes exclusively, whereas samples from Alabama to Texas mostly belonged to a second matrilineal assemblage.

In subsequent years, students in my laboratory were to extend such observations to other freshwater species in the region. For example, Kim Scribner examined mosquitofish populations across the Southeast for both mtDNA and allozyme variation. These cute little fish get their name from their habit of eating mosquito larvae. To our mammalian eyes, they have another endearing quality—internal pregnancy. Unlike most fish, which shed their gametes into the water and leave the parental duties strictly to Mother Nature, mosquitofish males have a penislike organ that they actively employ to impregnate a female, who, after a gestation period of three to four weeks, gives birth to several dozen live young.

Kim found that there were two genetically distinctive forms of mosquitofish (now considered separate species) in the Southeast. Again, one form occupied peninsular Florida and the Atlantic Coast, and the other occurred to the west. Joint inspection of the allozyme and mtDNA genotypes also revealed extensive hybridization between these species where their ranges overlap in western Georgia, Alabama, and eastern Mississippi. This geographic pattern of hybridization between distinctive genetic forms was mirrored closely in similar genetic studies of bluegill sunfish (assays conducted in my lab) and largemouth bass (work by others).

Another student, DeEtte Walker, assayed phylogeographic patterns in about ten species of aquatic turtles. These ranged from the adorable but odoriferous little stinkpots, musk turtles, and mud turtles, who like to bury themselves in stream or pond debris, to awesome snapping turtles, whose head and intimidating jaws alone can dwarf an entire stinkpot. Using nets and wire turtle traps, DeEtte collected turtle specimens from across the southeastern region, occasionally with my assistance.

DeEtte detected a variety of outcomes in her comparative molecular analyses of turtles, but several species showed deep phylogeographic subdivisions, much like those in the freshwater fishes. For example, two distinctive genetic groups of the loggerhead musk turtle were identified, one throughout Georgia and peninsular Florida, and the other in drainages to the north and west. Other species, such as the chicken turtle and pond slider, likewise showed pronounced regional differences in matrilineal phylogeny.

The genetic architecture of any species must be a joint product of contemporary and historical factors. Given the great diversity of molecular, ecological, and evolutionary agents impacting intraspecific genealogy over long periods of time, one might expect an idiosyncratic phylogeographic pattern for each species. Yet, our comparative molecular analyses of freshwater organisms in the southeastern United States were yielding recurrent themes. Furthermore, phylogeographic subdivisions within each species often aligned well with major biotic provinces as previously identified from range maps for hundreds of freshwater fishes and more than two dozen turtle species in the area. How could such apparent biogeographic order possibly have come to be?

One thing for sure is that relevant geographic and climatic factors in the region have been dynamic across recent evolutionary time. For example, periodic marine incursions over the last few tens of millions of years were largely responsible for demarcating today's flat, sandy coastal plain from the hilly, clay-rich piedmont. In Georgia, this "Fall Line" bisects the state along a path connecting Columbus, Macon, and Augusta, cities built where they are because the Chattahoochee, Ocmulgee, and Savannah Rivers yield hydroelectric power as they tumble onto the coastal plain, and navigable transportation as they then meander to the ocean. One prolonged marine incursion submerged an area now extending across the Okefenokee Swamp in southern Georgia and the Suwanee

River, cutting off a proto-Florida island, which today forms the central peninsular highlands.

At other times, ocean levels were lower than now, exposing large areas of continental shelf currently under the sea. This was particularly true during the fifteen or more glacial episodes of the Pleistocene epoch, each spanning roughly 100,000 years and separated by interglacial periods of about 10,000-year duration. During glacial advances, great volumes of water were frozen in ice sheets that buried much of North America and Europe, lowering global sea levels by as much as 150 meters. Then, an enlarged Floridian peninsula was separated by only a narrow sea channel from an expanded Cuba, which itself nearly touched a distended Yucatán Peninsula. Continental glaciers never reached the southeastern portion of North America, but their climatic and geologic influences certainly did.

All the while, freshwater drainages danced to the tune of landscape changes. For example, the Tennessee River now flows westward into the Mississippi, but it formerly drained southward through Alabama into Mobile Bay; and the Chattahoochee River, which flows to the Gulf of Mexico, was once in contact with the Atlantic's Savannah River when their headwaters touched in the southern Appalachians. In general, during periodic high-sea stands of the last few million years, coastal portions of most rivers were drowned, such that any connections between them would have been confined to their headwaters. Conversely, during low-sea-level stands, adjacent rivers occasionally merged nearer their mouths as they meandered across broader coastal plains.

Like passengers aboard a capricious transit system, freshwater populations must have been ferried about, sundered, sometimes merged, and often driven to extinction by such geophysical alterations of aquatic habitat. Species were more than hapless commuters, however, because their inherent dispersal abilities also would influence their distributions and genetic architectures. In general, we came to appreciate that the molecular data were doing their best to convey the genealogical tales of the recent evolutionary past. Our tasks became to translate these coded memoirs, interpret the genetic autobiographies of each species, and ultimately to reveal the composite sagas of regional biotas.

In particular, concordant phylogeographic patterns across multiple freshwater taxa probably evidenced shared histories of dispersal, molded in large part by the geophysical histories of the river drainages occupied. Our molecular data for numerous species indicated that aquatic populations in at least two southeastern regions had evolved independently for a considerable time, probably because of their confinement to long-isolated river basins. Superimposed on the deep phylogeographic accounts were more recent genetic entries into the evolutionary diaries. For example, the broad genetic hybrid zones in bluegill, mosquitofish, and largemouth bass probably were due to fairly recent secondary contacts, perhaps augmented by human-mediated transplantations within the last century.

Excited by this comparative phylogeographic approach, we widened our empirical work to the marine realm. A succession of graduate students devoted their time in my lab to examining molecular genetic patterns in a host of maritime creatures ranging from saltwater fishes, terrapins, and salt marsh birds to oysters and horseshoe crabs. Typically, we collected each species at coastal locales throughout the northern Gulf of Mexico, peninsular Florida, and along the Atlantic coast to New England. We also purchased a few samples from the Gulf Specimen Company, a small biological supply firm near Panacea, Florida, owned and operated by a most interesting fellow, Jack Rudloe.

One collecting trip was for the diamondback terrapin, *Malaclemys terrapin,* arguably the most beautiful of North American turtles. An inhabitant of salt marshes, this terrapin is clawed (not flippered) and has a shell elaborately etched with concentric designs. They eat horse mussels, using heavy crunching jaws. I and a new graduate student, Trip Lamb, headed off for the Gulf Coast

to collect terrapins for mtDNA analysis. Our first stop was in Panacea, where we hoped to get local collecting advice from Jack Rudloe, with whom I had spoken by phone several times over the years. Jack greeted us warmly, but when we told him of our terrapin project, he retreated to his office and slammed the door. What could we have said that upset him so? Did he object to genetic work on the diamondbacks for aesthetic reasons, or perhaps from conservation concerns? We knocked repeatedly on his door, but he emerged only briefly to point us to a book on his shelf that he assumed (incorrectly) we had read. It was a book he had authored: *Time of the Turtle*.

Back at our motel room that evening, Trip and I read the relevant chapters. According to Indian folklore and fishermen's tales, diamondback terrapins bring horrible luck to those who disturb them. The turtle's hex was feared by local boatmen, who scrupulously avoid *Malaclemys,* the *mal*evolent beast. Being a scientifically trained naturalist, Jack initially had scoffed at this superstition, but after traumatic personal experiences, he too was scared. Jack wanted no more of the turtle's curse, rationality be damned.

It had all started years earlier, when Jack received a telephone order for several terrapins. Although warned against the collecting effort by his boat captain, Jack needed the money and gave the matter little thought. Within a few minutes of catching dozens of specimens from an offshore spoil bank, a squall materialized from nowhere. Their boat was pushed hard aground, its keel damaged and its transmission disabled. They struggled back to port, but to add injury to insult, a turtle latched onto Jack's toe and almost chewed it off.

On the way to the freight office the next day to ship the terrapins to the client, Jack wrecked his car. Still, he remained skeptical of any causal connections until later that weekend, when the circulating pumps in his company's aquarium facility failed. A smelly, expensive fish die-off ensued. During the repair operation, the problem was discovered: a male terrapin had climbed into a water pipe, died, and clogged up the whole works. Jack's book went on to relate other such incidents, which had profoundly changed his attitude toward these herpetological demons of the salt marsh.

Thus, it was with some trepidation that Trip and I headed out the next morning to an offshore island to collect our own diamondbacks. We set up a series of gill nets in the shallows and returned the next day to assess our catch. Although four terrapins were in the nets, so too were dozens of stingrays. From a poison-laden dart at the base of its tail, a stingray can deliver an excruciating jab that usually requires a hospital visit. We spent several miserable hours cutting the tangled net from the rays' flailing tails. Miraculously, we avoided getting stung.

Our next terrapin stop was farther down the coast, at St. Petersburg. I stayed ashore as lookout while Trip and a third member of our collecting crew left the dock on a beautiful, sunny afternoon. At dark, when they should have been returning to port, a thunderstorm hit. Two hours passed, and just as I was about to call the Coast Guard, I heard the putt-putt of our outboard engine. It turns out that the howling storm had left my crew unscathed but capsized a pleasure boat nearby. Trip was towing that flooded boat and its shaken, drunken crew back to port. We halfheartedly convinced ourselves that there was no turtle hex after all, but, with our two dozen terrapins in a box on the back seat, you better believe we drove back to Athens very carefully!

The kinds of phylogeographic patterns revealed in our genetic assays of maritime species were reminiscent of motifs we had uncovered in the freshwater realm. Notably, several species had pronounced regional genetic differences, and most populations in the Gulf of Mexico were readily distinguished from those along the Atlantic Coast. This held true for the diamondback terrapins, toadfishes, black sea bass, horseshoe crabs, American oysters, and seaside sparrows. Similar patterns involving a deep Atlantic–Gulf genetic separation later were documented by other researchers in ribbed mussels, marsh killifish, a beach-dwelling species of tiger beetle, and several other maritime species.

Furthermore, in some of the previously mentioned invertebrate species with mobile larvae, genotypes normally characteristic of Gulf populations appear to have "leaked" into southeastern Florida. Many of the sea's inhabitants, like the staghorn corals, have such dispersive larvae. These may drift in the water column for weeks or months before settling down and transforming into adults, who thereafter may be sessile or sedentary. Thus, the genetic patterns in southeastern Florida might have arisen when water-borne larvae had been ferried out of the Gulf of Mexico by the Gulf Stream. This major current of warm water moves through the Florida Straits and hugs the coastline of southeastern Florida before veering offshore, eventually crossing the Atlantic to moderate Europe's climate.

The horseshoe crab, a foot-long arthropod with rounded shell and spearlike tail, is one such beast with a dispersive larval stage. We caught the adults by walking coastal shallows or finding pairs mating on tidal flats. Another species with dispersive larvae is the American oyster. Because of its abundance and the ease of collecting it, we also studied this species intensively.

My students found abrupt shifts in the genetic makeup of both species near Cape Canaveral in eastern Florida. Cape Canaveral is in a coastal region of ecological and zoogeographic transition between temperate and tropical-adapted maritime faunas. The shift in physical environment occurs in part because the Gulf Stream veers offshore in this area, replaced by cooler nearshore waters from the north. Thus, many maritime species of fish and invertebrates occur

either north or south of this region, but not both. For the American oyster and horseshoe crab, which *are* distributed continuously across the Cape region, the population genetic breaks registered by mtDNA coincided nicely with the boundary between these two distinctive zoogeographic areas.

Today, the Florida peninsula protrudes into subtropical waters, separating many temperate species into disjunct populations along the Atlantic Coast and the Gulf of Mexico. This contemporary physical barrier no doubt helps to maintain the genetic divergence we observed in some of the temperate species such as the black sea bass. However, the recurring phylogeographic patterns that we discovered in the maritime realm again must have reflected shared historical factors as well. What might these be?

With each Pleistocene glacial cycle, tens of thousands of years in total duration, opposing influences on species distributions were at work. During glacial advances, climatic cooling would have pushed temperate species southward, perhaps boosting opportunities for genetic contact between populations in the Atlantic and the Gulf of Mexico. Yet, at these times of lower sea level, the Florida peninsula was also larger and more arid than now, bordered by fewer estuaries and salt marshes, which are favored habitats of many coastal taxa. Thus, for some other maritime species, environmental conditions at glacial maxima actually may have promoted a separation of populations between the Atlantic and the Gulf.

During balmier periods, such as the present, temperate species with narrow thermal tolerances or lower dispersal capabilities probably were sundered into separate Atlantic and Gulf forms by warm waters surrounding a smaller Floridian peninsula. Yet, at these interglacial times, species with broad ecological tolerances or greater dispersal proclivities probably expanded out of the Gulf to regain contact with populations in the Atlantic. Such glacial cycles and their impacts on faunal distributions were episodic and frequent throughout the Pleistocene.

No single historical explanation could account for the phylogeographic patterns we observed in all species surveyed. Atlantic and Gulf populations of several maritime species, such as sturgeon and two species of marine catfish, showed little or no detectable phylogeographic separation. In other species, the magnitudes of genetic divergence between the Atlantic and Gulf populations varied considerably. In theoretical models, we showed that such heterogeneity was consistent with various combinations of factors, including cyclical environmental changes precipitated by glacial cycles, heterogeneity in molecular evolutionary rates across the taxa examined, and differences in the demographic histories of species, which would influence rates of historical matrilineal sorting.

These models implied that our genetic data had to be interpreted cautiously. In evolutionary biology, "just-so stories" (a phrase taken from Rudyard Kipling's tales) tend to be discounted when they amount to little more than imaginative storytelling. We did not want to fall into that trap. Yet the recurring phylogeographic themes we had unveiled and their plausible relationships to each species' ecology and lifestyle could not be ignored. Eventually, by integrating the genealogical data with coalescent theory, we were able to develop a broad conceptual framework involving principles of "genealogical concordance" that has become a cornerstone approach in phylogeography for interpreting empirical outcomes in a comparative context. This approach has helped to clarify the intertwined roles of historical geography, natural history, ecology, and historical demography on the contemporary population-genetic architectures of species.

As a follow-up to our comparative work on southeastern faunas, we wondered whether analogous patterns of phylogeographic concordance might be apparent in other regional biotas as well. So, Trip Lamb began a genetic assessment of desert creatures in the American Southwest. In the first species examined, the desert tortoise, populations east versus west of the Colorado River displayed a deep genealogical split in mtDNA. Trip next examined two lizard species in the region, the desert iguana and chuckwalla. To our chagrin, the mtDNA assays revealed little commonality across these species in the particular geographic distributions of the female family names. Nonetheless, each species had its own idiosyncratic story to tell.

Another fascinating creature examined in my lab was the American eel. Found in freshwater streams throughout eastern North America, these slimy, ropelike creatures make a remarkable journey. Upon reaching sexual maturity at two feet long and ten years old, eels migrate downstream to embark on an oceanic voyage to the Sargasso Sea, a tropical region of the west-central Atlantic Ocean. There they apparently spawn en masse in deep waters. Actually, spawning has never been observed directly but is deduced from plankton-net trawls, which capture young eel larvae only in the Sargasso Sea. Conventional wisdom was that spawning occurs more or less at random, and that oceanic currents transport the larvae back to North America and randomly distribute them to the dozens of streams from which their ancestors had emigrated a year or more earlier. If so, we reasoned that eels found in all streams should be genetically similar, representing random genetic draws from a single, well-mixed breeding pool in the Sargasso Sea.

The life-history pattern in the American eel is rare in the biological world, as is the logical prediction that freshwater populations of this species are genetically homogeneous. Impressed by the dramatic phylogeographic structures that we had seen in most other marine species, we thought we might make a scientific splash by documenting strong genetic differentiation in eels also and thereby refute orthodox wisdom about the random nature of adult mating and larval migration. So we assayed mtDNA in eels collected from Louisiana to Maine. To our surprise, the data supported prevailing views. Each sample of eels in North America, regardless of source, displayed a jumbled mix of female family names. Here was a glaring exception to the emerging generality that most species display pronounced population structure, but it was an exception that helped to prove a broader rule: each species' natural history and life cycle can greatly impact its phylogeographic pattern.

Under the lead of another of my graduate students, Marty Ball, some of our attention turned to phylogeographic patterns in various avian taxa. For example, we wondered whether species in high latitudes, which must have colonized their current ranges following the retreat of the latest Pleistocene glacier, might in general show less geographic population structure than their counterparts to the south. We also wondered how an explosive historical growth in population size might impact the structure in a mtDNA gene tree. We focused initially on the redwing blackbird, one of the most abundant birds in North America.

This highly mobile species nests continent-wide in marshes and wet pasturelands. Sure enough, the mtDNA patterns in specimens we obtained from across the United States and Canada proved consistent with the species' high mobility and its suspected recent evolutionary past. Thus, all surveyed populations were close genealogically, and the redwing's matrilineal tree showed a "starburst" pattern in which most mtDNA genotypes traced as if by explosion from a few ancestral family names. This made sense in terms of computer-based simulations predicting such a pattern for a species that had recently expanded from a population refugium.

Marty extended his studies to other widely distributed bird species. Some, such as the downy woodpecker, showed a shallow phylogeographic pattern (like that of the redwing blackbird), whereas others, such as the rufous-sided towhee and common yellowthroat, displayed at least one deep genetic subdivision within their respective ranges. As in some other taxonomic groups that we had surveyed, much genealogical idiosyncrasy (of interest in its own right) existed across these avian taxa.

Another avian species studied in my laboratory was the snow goose. These beautiful creatures with white plumage and black wing tips nest in discrete colonies across northern Canada, Alaska, and parts of Russia. From years of banding efforts, it was known that females usually return to nest at their natal site after wintering in the southern United States or Mexico. Thus, we reasoned, each rookery might display a unique set of female family names, which could then serve as natural genetic tags to supplement the metal and plastic tags used by ornithologists to decipher migration routes. Also, field observations had documented that males pair with females on mixed-colony wintering grounds. Therefore, we hypothesized, nuclear and mitochondrial genes might paint very different phylogeographic pictures because of the sex-specific mating patterns and migrational behaviors of this species.

To our surprise, particular mtDNA genotypes proved not to be confined to specific nesting colonies. At first thought, this appeared to contradict the irrefutable results of artificial tag returns. However, the contradiction dissolved upon further reflection, and a broader scientific lesson emerged. Direct observations of contemporary animal dispersal (for example, from artificial tag returns) can convey a false impression of population connectedness because they fail to address *historical* movements within a species. Conversely, geographic distributions of genetic markers, taken at face value, can provide a spurious impression of *contemporary* dispersal and gene flow because they retain a footprint of historical movement patterns that sometimes may differ from those of the present.

In snow geese, nesting colonies probably have tight matrilineal connections because of past female movement between rookeries, colony turnover via extinction and recolonization, or the joint establishment of high-latitude rookeries by pioneers from a small number of Pleistocene refugia. Such historical demographic events affecting gene distributions might be common on the temporal scale of centuries or millennia yet rare during the years or decades available to field naturalists. Thus, a secure understanding of population structure in any species requires a thorough integration of genetic information and field observations. Again, phylogeography was to become a bridging discipline by promoting increased communication between the traditionally disconnected fields of genetics and natural history.

In 1987 I published a review entitled "Intraspecific phylogeography: the mitochondrial DNA bridge between population genetics and systematics." It dealt with my lab's decade-long experience with geographic patterns of mtDNA variation in numerous animal groups. In this paper, my students and I coined the word "phylogeography," defined this new discipline, framed its purview, and proposed several phylogeographic hypotheses relating molecular genetic patterns to species' historical population demographies, ecologies, and natural histories.

After 1987 the field grew explosively, as evidenced by the fact that the number of scientific publications employing phylogeographic ideas and terminology roughly doubled in each two-year interval for at least the next sixteen years. Phylogeography's most important contributions have been to emphasize historical, nonequilibrium aspects of microevolutionary process; clarify formerly unappreciated connections between population demography and genealogy; and build new bridges between population genetics, ecology, and phylogenetic biology. Not bad for a field with humble roots in the curiosity-driven mtDNA analyses of a few field mice and pocket gophers!

13 Ancient Mariners

In 1985 a student of exceptional physical stature and a scientific vision to match showed up at my office. Brian Bowen wished to use the matrilineal markers provided by mtDNA to study migratory and other behaviors in marine turtles. I had contemplated a similar project years earlier but had been stymied by the bureaucratic hassle of procuring research permits for these endangered animals and by the logistic hurdles of sampling turtle colonies from around the world. However, pointing out such challenges to Brian was merely like waving a red flag before a bull.

I have had a succession of memorable characters come through my lab, but none more so than Brian. He was a fanatical Deadhead who traveled widely to attend Grateful Dead concerts, and sometimes he seemed to embrace a Jerry Garcia lifestyle himself. But Brian is also an avid naturalist with a keen eye for intriguing scientific questions. He is a lucid writer and poised public speaker. Brian is also socially adept, a trait that served him well in dealings with permit agents and scientifically illiterate government officials. It also helped with police, who sometimes accosted him on overseas collecting trips. In field kits, Brian carried a small bag of sucrose powder for making a buffer preservative for tissue samples. To an airport security guard or wildlife officer, it looked suspiciously like cocaine. Yet, Brian had an uncanny ability to defuse tense situations, and except once or twice, he managed to remain out of police detention.

A story illustrates Brian's style. One hot summer night, we were camped on the outer banks of Georgia's Cumberland Island, monitoring the nesting behavior of loggerhead turtles. One member of our team had been patrolling a beach to the north in a dune buggy and raced back to report that a thunderstorm was headed our way. There was little time, and no cover. The storm settled over us within minutes. For the next hour, bolts of lightning rained down, exploding like fiery cannonballs. I lay face down in the sand, as terrified by nature's actions as I have ever been. When I peered up into the melee, which was illuminated as if by a gigantic strobe, there was Brian, standing atop a dune like a captain at the helm, chugging shots of tequila and saluting the heavens in appreciative toasts. In short, Brian approaches life with gusto.

There was other electricity in the air that evening. Steve Karl was another of my students who was to assume some of the genetic research on sea turtles for part of his dissertation. As the rain poured and thunder clapped on that fateful night, Steve met the woman of his dreams, another member of the field crew. Michelle and Steve married a few years later.

Seven species of sea turtles exist in the world today, all listed as endangered or threatened under the U.S. Endangered Species Act or by the World Conservation Union. Their decline has been precipitous. For example, about fifty million adult green turtles inhabited the Caribbean when Christopher Columbus sailed into the region five centuries ago. His shipboard diaries tell of frequent collisions with these animals, causing consternation among his crew. Within a short time, many nesting colonies, such as those in the Cayman Islands, were extirpated as humans slaughtered the adult turtles for meat and harvested their eggs. Only a few hundred thousand green turtles remain in the Caribbean region now, about 1 percent of their pre-Columbian numbers.

Turtle meat is delicious, and sailors often stored live adults aboard their ships during long voyages. To set these stores, a ship's crew would wait at a turtle rookery for nesting females to come ashore, and then, as noted in one traveler's diary from 1691,

> turn them by surprisal on their backs, which is a posture they are utterly unable to recover from, and are thereby frustrated of all defence or escape, and are a ready prey to any that resolve to seize them. When the sensible creatures find themselves in this desperate posture, by which they know themselves to be in a lost and hopeless state, they then begin to lament their condition in many heavy sighs, and mournful groans, and shed abundance of water from their eyes, in hopes, if possible, to secure their safety by their tears, and mollifie the cruel assaults upon their lives. (Quoted by Roger Huxley, *Maritime Turtle Newsletter* 84 [1999]: 7–9, in his review of historical documents on maritime turtle exploitation.)

Other sea turtle species declined precipitously, too, notably the hawksbill, whose glossy carapace is used to make "tortoiseshell" jewelry. In response to conservation concerns, the National Research Council (an advisory arm of the National Academy of Sciences) recently published a book documenting human sources of turtle mortality. These include accidental drownings in commercial shrimp trawls and incidental "by-catch" in longline and drift-net fisheries. An even greater threat is the loss of nesting habitat. Real estate developers covet sandy beaches that the animals require for nesting.

Laying eggs ashore is a perilous event in a turtle's life. Consider the green turtle. In the dark of night, a gravid female lumbers out of the surf and hauls herself to the edge of the dunes. Using her flippers, she excavates a pit into which she deposits 100 or more eggs, each resembling a wet ping-pong ball. Covering the hole with sand, she returns to the sea, never to see her offspring. Eight weeks later, the babies hatch and dig themselves free from the underground womb. Their scramble across the beach must be quick, because gulls, raccoons, and other predators take a heavy toll. So too do sharks, barracudas, and other fish lurking in the shallows. Each defenseless hatchling must paddle offshore quickly, because its best hope for survival is to be swallowed by the vastness of the sea.

Where the youngsters travel for the next several years remains mostly a mystery. Perhaps they congregate along the boundaries of ocean currents, where floating rafts of sargassum provide food and refuge. Young green turtles adrift in the surface gyres may circumnavigate an entire tropical ocean basin before settling along a shallow continental margin, where they switch to a diet of seagrass. There, they reach sexual maturity at about the same age as humans and live out a comparable lifespan of many decades.

Turtles mate at sea, rather clumsily, after which gravid females return ashore to nest, usually several times per summer on a two- or three-year cycle. The nesting beaches are often far from the feeding pastures, meaning that females must migrate hundreds or even thousands of miles. In terms of the navigational skills and energy required, these migrations rank among the most impressive of such feats displayed by any creatures (birds, eels, and salmon included).

Some female green turtles nest on Ascension Island, one of the most isolated specks of land on Earth. This volcanic cone, six miles in diameter, is between Brazil, 1,400 miles to the west, and Angola, 1,900 miles to the east. The nearest terra firma in any direction is St. Helena Island, 900 miles southeast. Ascension rises from the Mid-Atlantic Ridge, an otherwise undersea mountain range. Green turtles nest in abundance on its dozen sandy beaches after a journey of several weeks from pastures along the South American coast. After laying eggs over a two-month period, the females complete their 2,800-mile odyssey by returning to Brazil.

These facts were discovered largely through the field efforts of the late Pro-

The primary nesting beach for green turtles on Ascension Island.

fessor Archie Carr. Throughout his career, this beloved naturalist probed the mysteries of sea turtle life. One approach by Carr and his students was to secure coded tags to the flippers of nesting females at their rookeries, Ascension Island included. Subsequent tag recoveries revealed the migratory routes of particular individuals. Cumulatively, the data led to an impressive conclusion. Throughout her life, an adult female typically remains faithful to one nesting beach. For example, nearly 30,000 females were tagged at the Tortuguero colony in Costa Rica, but only one was found nesting at any other monitored rookery in subsequent years.

Archie Carr was especially intrigued by the Ascension migratory circuit. Why would turtles undertake an incredible journey from South America to nest in this improbable locale? It made little sense, because suitable nesting beaches used by other green turtles were situated in Brazil and elsewhere along the South American coast. To explain this mystery, Carr published a bold hypothesis in 1974, when the phenomena of plate tectonics and continental drift had first gained convincing empirical support.

As anyone examining the jigsaw puzzle of continental plates on a world map might have surmised, the Earth's landmasses have not always been in their present configuration. For example, nearly 100 million years ago, the equatorial bulge of South America and the corresponding armpit of Africa were snuggled against one another, separated only by a narrow sea channel. Over geological time, these continents drifted slowly apart as volcanic action on the Mid-Atlantic Ridge spewed undersea lava, which caused the seafloor to spread by an inch or two per year. Like slow-motion bumper cars, all of the continents have floated about the Earth's surface on such geophysical conveyors, colliding, separating, and ferrying around their biotic passengers.

Carr's hypothesis of how green turtles came to inhabit Ascension Island appealed to a chain of events beginning some sixty to eighty million years ago. Then, he suggested, a proto-Ascension lying between South America and Africa was colonized by turtles from feeding pastures along the nearby continental margin. Carr had been impressed by the tag data documenting a strong fidelity of each adult female to a particular rookery. From those observations, he made an inferential leap—that each female was faithful to her natal rookery, the site where she herself had hatched decades earlier. Coupling this "natal homing" idea with the geologic evidence for continental drift, Carr concluded that over tens of millions of years, females nesting on Ascension had gradually lengthened their migratory circuit inch by inch, generation after generation. From the perspective of a land-based observer in South America, Ascension Island marched slowly seaward, and presumably the turtles followed its progression.

Yet there was no direct evidence of natal homing in marine turtles, for one simple reason. No one had ever devised a physical tag that could be applied to a silver-dollar-sized hatchling turtle and that would last for decades in the caustic marine environment to be retrieved from a quarter-ton adult. There were other reasons to be skeptical of Carr's scenario, not the least of which is that it requires the continual presence of suitable nesting habitat on Ascension Island for more than 60 million years. If the island's fragile beaches were destroyed for even one turtle generation across that sweep of time, the matrilineal circuit would have been broken. Such a break in one or more links of the migratory chain would seem inevitable. Today, for example, a rise in sea level by even a few meters would flood Ascension's few sandy beaches. Volcanic eruptions on Ascension, too, the last of which occurred 2,000 years ago, could at any time have severed the intergenerational chain of successful migration.

To Brian and me, this seemed an ideal evolutionary stage for mtDNA analysis. We thought that if the Ascension rookery truly was colonized more than sixty million years ago, and its matrilines were isolated ever since via female natal homing, then the female family names of Ascension nesters should be highly distinct from those of all other green turtles. This expectation held regardless of how many nuclear genes were exchanged between colonies (if Ascension females mated with foreign males). Alternatively, if Ascension was

Members of the turtle-collecting crew on Ascension Island: (from left to right) Brian Bowen, Ruth Bowen, and the author.

colonized recently by female turtles hatched elsewhere, there would be close matrilineal ties to the emigrant site.

We had a compelling research question, so now we needed to collect turtle eggs from Ascension and elsewhere for genetic analysis. Ascension Island is militarily strategic, having been on a supply route for the Allied effort in North Africa during World War II and the British operation in the Falklands War many years later. There are no indigenous peoples, but the island houses about 1,400 military personnel associated with the U.S. and U.K. Air Force bases. Except for a few private contractors, essentially the only civilians on Ascension are its British administrator and his staff.

After a year of letter writing and inquiry, Brian finally worked his way up the chain of command to the U.S. Assistant Secretary of Defense, who issued us official clearance to visit Ascension. Step B involved writing a grant proposal to pay for our seats aboard a military transport plane that flew to Ascension weekly out of Florida. Fortunately, the National Geographic Society came through with funding.

Another step was perhaps the hardest: getting research permits. Marine turtles fall under the jurisdiction of CITES, the Convention on International Trade in Endangered Species. Federal and state permits must be obtained to import turtle eggs into the United States, and wildlife officials greeting our return would require formal permission documents from the exporting nation as well. After months of effort, all authorizations were obtained except one. Despite many inquiries, we couldn't identify an office in England or Ascension with jurisdiction for an export permit. Thus it was with some concern that we departed Florida, uncertain whether we would be able to obtain an official export license from Ascension's British administrator.

The turtle crew consisted of me, Steve, Brian, and Brian's wife, Ruth, also a turtle biologist. Ascension Island is otherworldly. The first impression is that of a small lunar landscape, stark and austere, lost in oceanic space. Except for the towering peak of Green Mountain, the tortured landscape is bare volcanic rock in hues of red and black, virtually devoid of vegetation. Even the military's golf course follows this theme. The fairways, wending between hillocks of old lava flows, are of cinder and volcanic ash, and the "greens" are of pressed beach sand.

No visitor to this lonely outpost remains anonymous. We were greeted by the U.S. base commander, who, after arranging our lodging in barracks, asked whether I would deliver a talk at some point describing our turtle work to the Ascension community. I happily agreed. We were also introduced to Phillip Ashmole and his small research crew. I knew this ornithologist's name from the scientific literature. Thirty years earlier, Dr. Ashmole had led a British expedition to Ascension Island to describe its native birds.

In prior centuries, predator-free Ascension Island had been a prime nesting site for seabirds in the tropical Atlantic. Reliable tales were told of dense, clamorous breeding colonies stretching as far as the eye could see across the island's extensive lava fields. Sadly, most of the rookeries later were decimated by imported goats and cats. The intent of Phillip's earlier expedition, like his current effort, was to census the breeding birds to aid in conservation plans.

With the exception of a few sooty tern colonies, all of the nesting now occurs solely on Boatswain Island, a tiny satellite separated from Ascension by a sea channel just a few hundred yards wide. Steep cliffs and sharks in the strait have conspired to keep Boatswain free of feral cats, and about a dozen seabird species nest in local abundance on its ledges and across its flat top, including tropicbirds, black noddys, an endemic frigatebird, three species of booby, and the snowy-white Atlantic fairy tern.

Given our concern about getting export permits for turtle samples, we were dismayed to learn from Phillip that his research team had been denied access to Boatswain. Ascension's British administrator, a self-proclaimed conservationist by the name of J. J. Beale, had been hostile to Dr. Ashmole's biological research on the grounds that it might disturb the wildlife. We were later to learn that this was a rather disingenuous stance, because Mr. Beale, an avid cat lover, was opposed to environmentalist efforts to control the burgeoning number of feral cats on Ascension. These marauders devastate the island's fauna by eating the eggs and hatchlings of native seabirds and turtles. They also feed on Ascension's endemic scorpions, crabs, and a handful of other invertebrate species, whose distant ancestors somehow had made their own improbable way to this isolated biological outpost.

One night we were invited to a formal dinner at Mr. Beale's house, a spectacular mansion halfway up cloud-topped Green Mountain. Phillip and his wife were invited as well. What a tense evening that was! The Ashmoles could barely contain their bitterness toward the administrator who, by denying access to Boatswain, had compromised their expedition's mission to reassess Ascension's jeopardized avifauna.

Over cocktails, a simmering debate erupted into harsh words. The Ashmoles favored efforts to preserve Ascension's unique native fauna, which had evolved in isolation over thousands or millions of years. Mr. Beale's stance was that most of the feral cats now alive on the island were born there as well, albeit from recent domestic escapees, and, thus, had just as much right to the place. We were in total sympathy with the Ashmoles, yet felt compelled to feign neutrality lest the administrator likewise destroy our mission by denying the export license for turtles. All the while, Mr. Beale's cats swarmed the verandah and mansion grounds.

Like helpless defendants awaiting sentencing, we sweated through the cocktail hour and the heated conversation at the dinner table. Afterwards, over liqueurs, Brian came to our rescue. Mr. Beale clearly enjoyed Brian's relaxed conversational style and, with the tension lifted, invited us for a round of croquet. On the lighted lawn, I made several shots that Mr. Beale deemed jolly good and the sure mark of a cultured upbringing. By evening's end, Mr. Beale proclaimed that he would grant us the turtle permits!

Greatly relieved, we soon were back in the more comfortable abode of nature. Pan-Am Beach, near the air base, is where many of Ascension's turtles nest. About a quarter-mile long and flanked by lava hills, this sandy shore became our center of operations for two weeks as we monitored turtle nesting activities from dusk to dawn. Each night, we stationed ourselves strategically along the beach to sample from as many nests as possible.

Lying prone in the sand, we scanned the moonlit surf to spot each dark form emerging from the sea. Ascension's green turtles are among the largest in the world, weighing several hundred pounds and approaching five feet long. Like a shadowy prehistoric monster, each behemoth worked her way slowly up the beach, sometimes passing within a few feet and showering one of us with sand from her great flippers. We had to lie perfectly still lest she became alarmed and returned to the sea. As she continued up the beach, one of us would crawl silently on our belly thirty feet behind, hunkering down where possible into abandoned craters dug by earlier nesters. The beach resembled an artillery range pockmarked with such pits, because each female often digs several holes before laying eggs.

Finding a suitable spot, the turtle then spent upwards of forty minutes excavating a four-foot-deep depression using rhythmic thrusts of her front and rear flippers. A female entered the final egg-laying phase shortly after she scooped out a basketball-sized cavity in the bottom of the pit and rigidly splayed her hind feet over it to keep the sand from caving in.

At that stage, a female's wariness is lost completely as she enters a quiescent state. She could be approached with impunity now, and indeed a marching band would not alter her behavior. Crawling down into the hole, we reached between her hind flippers to catch an egg as it was being deposited. Removing one or two of these conveniently packaged DNA samples does no harm to turtle reproduction. Each female lays 100 or more eggs in the next few minutes, several hundred more during the season, and many thousands across her lifetime. Yet from all of those eggs, only about two per female on average result in offspring that will survive to adulthood. Her job completed, each female then covered her nest with sand and hauled herself back to the sea. On a good night, each member of our crew got samples from three or four nests.

Those evocative nights on Ascension filled my senses with lasting, romantic impressions: the gentle warmth of humid breezes off a tropical ocean; the abrasive grit of volcanic sand on arms and legs; the shadows of contoured hills against a moonlit sky; the sounds of a pounding surf in the background; and the heaving sighs of giant turtles. We were transient visitors on a tiny speck of land lost in space and time, deeply honored to be in the company of such extraordinary animals, who had traveled an incredible distance to complete a mission far more important than our own. As I peered out into the sweeping black sea from which the turtles had emerged, and then up into the panoramic heavens studded with stars, I felt a special attunement to the universe and to all of Creation.

Later, our genetic assays revealed the colonization history of the Ascension rookery. Dr. Carr's provocative hypothesis of an ancient origin linked to plate tectonics proved incorrect. Based on the close matrilineal similarity of Ascension females to those nesting along the coast of South America, the island must have been colonized within the last 100,000 years at most. As judged by genetic and geographic evidence, the current nesting colony on Ascension began when one or a few gravid waifs recently colonized the site from continental sources.

Nonetheless, the female family names carried by Ascension nesters differed consistently, albeit at a low level of genetic divergence, from those of green turtles at some other South American rookeries. From tag returns, Archie Carr knew, for example, that females from Suriname and Ascension shared feeding areas in eastern Brazil. These were important pieces of evidence with regard to the broader hypothesis of natal homing. When turtles share feeding grounds but differ rookery by rookery with respect to mtDNA genotype, it must mean that females normally return to nest at their respective natal sites.

To appreciate why this is so, consider an alternative scenario, which also would be consistent with the well-documented fidelity of *adult* females to a particular nesting locale. Under the "social facilitation" model, a first-time breeder follows an experienced female from feeding areas to an established rookery, and after a successful experience "fixes" upon that site for all subsequent nesting throughout her lifetime. However, the social facilitation hypothesis predicts that rookeries should often share matrilines when feeding grounds overlap, because an inexperienced female might follow an experienced nester to a foreign site. For Ascension and Suriname, which shared no mtDNA genotypes, the molecular data appeared more consistent with the natal homing scenario. This conclusion was later to gain support as we assayed additional rookeries around the world.

Not all of our time at Ascension was devoted to turtle activities. In an old truck loaned to us by a construction crew, we explored the rain-soaked summit of Green Mountain, toured ancient lava beds and rock formations at lower elevations, and visited the settlement of Georgetown. Standing on a cement pier jutting into the town harbor, we were amazed by the clouds of black triggerfish swarming below. This species, not particularly common elsewhere in the Atlantic, is abundant in Ascension waters and, in piranha-like fashion, serves as the community's garbage disposal service.

The Air Force maintained a SCUBA boat as part of its recreational program, and we boarded it for an extraordinary dive into the pristine waters surrounding Boatswain. There, below towering rocky cliffs, we dove amid shoals of triggerfish and other species, including creoles, soapfish, squirrelfish, sergeant majors, damselfish, gobies, surgeonfish, sharks, and moray eels. As we surfaced, we were engulfed by hundreds of seabirds diving and reeling about. With clouds of fish below and birds above, I felt like an integral chunk of a rich biological chowder.

During their graduate tenures, Brian and Steve collected additional turtle samples from around the world. I wish I could have accompanied them to Suriname, Greece, Australia, Mexico, Costa Rica, and elsewhere, but someone had to stay home to deliver lectures, write papers, review grant proposals, edit journals, answer correspondence, attend meetings, and in general perform the normal tasks that in truth occupy the great majority of an academician's life! In global surveys of the green turtle, Brian found a relatively deep genetic schism distinguishing rookeries in the Atlantic and Mediterranean from those in the Pacific and Indian Oceans. This made considerable sense, because the species probably was split into these two distinctive evolutionary units about three million years ago, when the Isthmus of Panama arose and first provided a terrestrial barrier between tropical marine populations in the Atlantic and Pacific.

Oddly, another group of marine turtles that we genetically assayed should have been affected similarly by the Panamanian land bridge yet failed to mirror the green turtle's global phylogeographic pattern. Traditional wisdom says that Ridley turtles come in two forms: the olive ridley, found throughout large portions of the Indian, Pacific, and Atlantic Oceans; and the highly endangered Kemp's ridley, confined to a single major nesting population in the Mexican state of Tamaulipas along the Gulf of Mexico. However, according to Archie Carr, these distributions "make no sense at all under modern conditions of climate and geography." This conundrum, together with the close morphological similarity between the Kemp's and olive ridleys, had raised doubts about their taxonomic recognition as two distinct species.

Earlier in this century, Kemp's ridleys nested on the Tamaulipas beach by the tens of thousands in a dazzling *arribada* (Spanish for "arrival"). During daylight hours, females came ashore en masse, laying millions of eggs that would hatch synchronously several weeks later. Perhaps the arribada reduced hatchling mortality by supersaturating terrestrial and marine predators. In any event, this nesting behavior also made the rookery susceptible to human exploitation, and by the latter half of the twentieth century, only a few hundred members of the species remained. On the verge of extinction, the Kemp's ridley finally gained conservation attention, and international recovery efforts were launched. These involved attempts to protect the original nesting site and establish a new rookery in Texas by transplanting eggs and then "head starting" the hatchlings (rearing them in captivity until later release).

To address whether these extraordinary recovery efforts were directed toward a "valid" species, we compared mtDNA genotypes in the Kemp's ridley with those of olive ridleys from the Pacific and Atlantic. Expecting to find that Kemp's was merely a genetic garden variety of the olive ridley, we were surprised to discover in the matrilineal tree a deep subdivision that matched the traditional species' designations perfectly.

In 1969 another turtle researcher, Peter Pritchard, had advanced a biogeographic hypothesis for the ridley complex that now made excellent sense in the light of our genetic findings. The basic idea was that long ago, an ancestral ridley population was split by the Panamanian isthmus into proto-Kemp's and proto-olive forms in the Atlantic and Indo-Pacific basins, respectively. Much later, olive ridleys from the Indian Ocean recolonized the Atlantic via Africa's Cape of Good Hope, thus accounting for the species' current distributions. This speculative scenario now had empirical support. Furthermore, our genetic findings bolstered the taxonomic assignments upon which conservation efforts for the Kemp's ridley had rested.

There were other conservation-relevant aspects to our genetic data on various sea turtle species. Although green turtle rookeries within an ocean basin were related closely, they often displayed slightly different mtDNA family names, an observation generally consistent with natal homing by nesting females. In turn, natal homing suggests that most rookeries are demographically autonomous with regard to female reproduction, at least over time scales germane to conservation efforts, because if most females return home to their natal site, a rookery that goes extinct is unlikely to be reestablished over ecological time by females hatched elsewhere. These genetic inferences are consistent with empirical experience. Several rookeries (such as on Cayman Island) that were extirpated by humans in the past 400 years have yet to be recolonized naturally.

Of course, natal homing cannot be absolute, because multiple rookeries exist. Turtle rookeries within an ocean basin often displayed distinguishable mtDNA genotypes, but their close matrilineal affinities implied that they had evolved roughly in genetic concert over deeper time. Our suspicion is that rookeries are ephemeral, blinking off via occasional extinction and sometimes coming back to life when recolonized by gravid strays hatched elsewhere. Thus, overall, there appears to be a strong tendency for natal homing by females, coupled with rookery turnover in short evolutionary time.

The identification of rookery-specific mtDNA markers also enabled Brian to identify the natal origins of marine turtles on feeding grounds and migration routes. For example, he used our matrilineal markers to show that loggerhead juveniles hatched in Japan apparently traverse 6,000 miles of open Pacific to utilize feeding arenas along the coast of Baja California. In another such study, about 50 percent of the loggerhead juveniles killed in a longline fishery in the Mediterranean Sea proved by genetic evidence to have been hatched on nesting beaches in the Americas. In general, marine turtles captured on particular feeding grounds or migration routes often proved to have originated from two or more rookery sites. It logically follows that any local mortality factors at non-nesting phases of the turtle life cycle can negatively impact multiple rookeries. This too has management implications.

The genetic findings on marine turtles also raised jurisdictional and legal issues. For example, the *United Nations Convention on the Law of the Sea,* to which more than 150 countries are signatory, prescribes that the state which provides developmental habitat for a species holds primary responsibility for conserving that stock. This law was written with migratory marine fishes in mind but clearly can apply to sea turtles as well. When turtles are killed on the high seas (for example, in longline fisheries), shouldn't a country that provided nesting habitat for these endangered animals have legal and moral rights to demand conservation agreements with the harvesting nation?

We also employed the molecular data to reconstruct a phylogenetic tree for all marine turtle species and compare it to previous estimates based on morphology and a fossil record that extends back 150 million years. One interesting question concerned the evolutionary origin of the hawksbill turtle's unusual diet. This species eats sponges almost exclusively. Did sponge-eating evolve from an ancestral condition of plant-eating (the feeding habit of adult green turtles) or from meat-eating (the custom of other marine turtles)? Based on the hawskbill's placement in the phylogenetic tree, this species' ancestor was most likely a carnivore.

Our first genetic paper on marine turtles was published in 1989, and it initiated a tidal wave of interest in the topic. When the literature was reviewed a scant decade later, more than 120 research papers involving many laboratories had dealt with the population genetics, phylogeography, molecular evolution, and conservation of these incredible seafarers.

Marine biologists have always taken pride in devising ingenious methods to delve into the many hidden facets of a sea turtle's life. Growth rates and ages of individuals are read from patterns of bone deposition, reproductive histories of females are deduced by laproscopic examination of oviducts, and movement patterns are chronicled from tag recoveries as well as from the composition of barnacles and other small hitchhikers attached to a turtle's shell. Our genetic research on marine turtles carried forward this tradition of scientific sleuthing, and it exemplified nicely the theme of my broader research program. In this case, each piece of genetic information gleaned about the incredible lives and evolutionary histories of the ancient turtle mariners was hard won, and we felt privileged that nature would share with us such precious secrets.

14 Species and Their Conservation

Nature's astounding variety, so emotionally uplifting to many of us, is displayed abundantly to those who take the time to look. Another sort of attachment comes from probing her deeper intellectual mysteries—how nature came to be what she is. This is the task of the evolutionary geneticist, to probe below the surface and into the past to uncover the mechanistic processes and forces that have molded biotic diversity. One such topic that has occupied my attention is how new species arise.

What is a species? Field guides tend to oversimplify the matter by portraying idealized specimens, often with arrows pointing to diagnostic morphological traits. The notion that members of a species share particular defining features is referred to as the "nondimensional" (in time and space) concept of the local naturalist. In practice, it works well in partitioning the biotic world into seemingly discontinuous local units that sometimes are perceived consistently by independent cultures. For example, the preliterate highlanders of New Guinea have vernacular names for 136 of the 137 native bird species recognized by Harvard-trained ornithologists. Likewise, native Indians and academic scientists recognize basically the same set of tree species in Amazonia.

However, such agreement begs the issue of why members of a species display uniting characteristics. For most of the twentieth century, conventional scientific wisdom held that conspecific populations share traits because they belong to the same gene pool. If organisms interbreed, their genes in effect are mixed together during sexual reproduction. Alternatively, if they are reproductively isolated by genetic differences in mating habits or by an inability to produce viable and fertile offspring, their gene pools are likely to have diverged over time. This in essence is the popular biological species concept (BSC). Decades ago, the evolutionist Ernst Mayr defined biological species as groups of "actually or potentially interbreeding natural populations which are reproductively isolated from other such groups."

However, this abstraction is not particularly user-friendly. First, it is often difficult to directly observe mating events and their outcomes in nature, so surrogate data (typically from diagnostic morphological traits, such as plumage color) are employed instead to deduce reproductive boundaries and demarcate species. Second, it usually remains uncertain whether a suspected potential to interbreed might be realized if allopatric populations (those occurring in different geographic areas) were afforded the opportunity in nature. Third, in some cases reproductive isolation may be partial rather than complete, such that shades of gray can exist along the black-to-white continuum between conspecific populations and well-demarcated species. The net effect of these practical difficulties is that educated guesses and subjective judgments sometimes are required about precisely where to draw species lines under the BSC, particularly when dealing with allopatric populations.

In response to these and other difficulties, a competing view has gained popularity in recent years. According to the phylogenetic species concept (PSC), a species is any population or group of populations phylogenetically differentiable from others, regardless of reproductive compatibility. The most vocal of the PSC proponents make explicit their distaste for the BSC by stating flatly that reproductive isolation should not be a part of species concepts. Yet the PSC also has shortcomings regarding how phylogenetic discontinuities are to be recognized and ranked. Indeed, each family group and even individual is genetically distinct at a high level of resolution, but there is no point in naming all such entities different species.

Part of the difficulty of characterizing a "species" is that the word is forced to serve at least two masters: the taxonomist, who merely seeks a utilitarian nomenclatural scheme, and the evolutionist, who desires that the organic entities named reflect the biological outcomes of natural processes (such as the erection of reproductive isolating barriers under the BSC, or phylogenetic splits under the PSC). As my academic grandfather Dobzhansky noted seventy years ago, "Biological classification is simultaneously a man-made system of pigeonholes devised for the pragmatic purpose of recording observations in a convenient manner and an acknowledgment of the fact of organic discontinuity."

My own interest in speciation processes began in graduate school but later was greatly reoriented and invigorated by the unorthodox perspectives demanded by mtDNA. Under the BSC, a sexual species is a reproductive community whose gene pool retains coherency via interbreeding between males and females. In a sense, conspecific populations are knotted together by the "horizontal" ties of matrimony. Yet, the "blessedly celibate" mtDNA molecule is transmitted asexually (the phrase "blessedly celibate" is from Richard Dawkins's *River out of Eden*). Thus, any genetic ties among matrilines must be strictly "vertical," united only by virtue of tracing to shared female ancestors. In attempting to account for the biotic discontinuities in nature that we call species, I wondered how the vertical bonds of phylogenetic ancestry might be reconciled with the horizontal bonds of contemporary interbreeding.

When I first thought about these issues twenty-five years ago, it seemed unlikely that asexual mtDNA genotypes might be woven into phylogenetic braids that generally would match, in theory or in practice, biological species as defined by reproductive criteria. Yet, in our genetic assays of rodents, birds, turtles, lizards, fishes, and other vertebrates, each taxonomic species usually did seem to have a coherency with respect to its maternal phylogeny. This often proved true even for closely related forms. For example, in one study we examined mtDNA genotypes in several pairs of avian species so similar in morphology that even experienced birders can barely distinguish them in the field. Nevertheless, king and clapper rails proved readily separable in mtDNA genotype, as did long-billed and short-billed dowitchers, and boat-tailed and great-tailed grackles. In the mtDNA assays, these and several other pairs of related species stood in strong distinction from one another.

In reviewing many such examples from my laboratory and the broader literature, we were led in one paper to conclude that molecular phylogeneticists could join the New Guinean highlanders and Amazonian natives, as well as traditional academic systematists, with respect to our shared perceptions of biological discontinuities (species) in nature. Why might such discontinuities exist and often be consistently recognizable? I now believe that the answers require an explicit focus on historical population demography (in addition to reproductive relationships per se).

To appreciate the ideas, even in simple terms, requires an introduction to coalescent theory. The word "coalescent" refers to the fact that if one could peer into the past, all extant lineages within a species eventually would trace or coalesce to common ancestors. For example, all human matrilines alive today ultimately stem from one great, great, great . . . grandmother of us all. Scientists as well as the popular press dubbed her Eve, the mother of all matriarchs, and extensive mtDNA data suggest that she lived about 200,000 years ago in Africa. However, Eve was not alone. As pointed out by the often counterintuitive niceties of mathematical coalescent theory, she was probably merely one of about 20,000 individuals alive at that time, many of whom undoubtedly contributed to our endowment of nuclear (but not mitochondrial) genes.

Likewise, every species must have its matrilineal Eve. Consider the only hypothetical situation under which this would not be true. Suppose that in every generation, each adult female produced exactly one surviving daughter, no more and no less. Then, all matrilines would survive forever as a vertical series of parallel lineages, never joining, never coalescing. There would be no unique Eve in a species and indeed no phylogenetic tree of matrilines. Instead, there would be a constellation of Eves, exactly as many as there are living females. Of course, no real species behaves this way.

At the other end of the spectrum, consider a hypothetical species where in each generation only one adult female is the biological mother of *all* surviving daughters. No other mothers contribute to the daughter pool. Then, looking backward at any point in time, all extant matrilines would have coalesced exactly one generation earlier. In each generation, all living twigs in the species' matrilineal tree would have burst forth from a single bud. Of course, no real species behaves for long in this way either.

In the real world, females in any generation produce varying numbers of surviving daughters, some more and some less. This is why we can speak of a matrilineal phylogeny for a species and draw meaningful analogies to a real tree, whose branches connect at historical nodes and ultimately trace back to a common root. Any phylogenetic tree of matrilines self-prunes in each generation. Some twigs inevitably are lost even as others may take their place by sprouting from the more successful matriarchal buds. This process is known as "lineage sorting."

One task of coalescent theory is to estimate the evolutionary age of shared ancestors as a function of a population's demographic history. For example, in a hypothetical, stable-sized population of 10,000 adult females, coalescent theory shows that all extant matrilines likely trace to an Eve who lived about 20,000 generations earlier. Such a recent date may seem counterintuitive, but it's the correct statistical expectation. Mathematical theory is most enlightening precisely when it challenges intuition.

Most populations in nature fluctuate in size and, hence, violate an important assumption of the foregoing scenario. For example, about six billion people crowd the planet today, but the total human population size was vastly smaller even a few centuries ago. Demographic complications of this type can be accommodated under coalescent theory by the concept of "effective population size" (N_e), which refers to how many individuals would comprise an idealized population that behaves genetically like the simplified ideal. An N_e value itself can be calculated (and the time of Eve estimated) if historical demographic factors are specified.

With help over the years from some mathematically gifted graduate students and colleagues, I have conducted many studies attempting to meld coalescent theory with species concepts and speciation theory. Our goal has been to integrate coalescent ideas with molecular data on mitochondrial (and nuclear) gene phylogenies to provide fresh insights into why the biological world appears to be subdivided into units that we typically recognize as species.

First, I have come to believe that most species are, in effect, small in breeding population size over extended evolutionary time. For example, adult American eels now number in the millions, yet in molecular-genetic properties the species appears as if its long-term population size in effect was only about 6,000 breeding females per generation, on average. Similarly, redwing blackbirds today number at least forty million adults, yet the genealogical features of the species (deduced from mtDNA) indicate a one-thousand-fold lower evolutionary N_e. Such findings suggest that most species fluctuate greatly in abundance and/or that breeding individuals produce highly variable numbers of offspring. In either case, gene lineages get "squeezed" through fewer ancestors than otherwise might be supposed. Such molecular-genetic signatures of past demographic contractions have been discovered in many now-abundant species, ranging from marine copepods to humans. Again, the genetic data evidence the dynamic, nonequilibrium nature of nature.

A second revelation from our studies is that each gene in a sexually reproducing species has a different genealogical story. Mitochondrial DNA records the history of female lineages, but each gene in a cell's nucleus has had its own idiosyncratic history of transmission through male and female ancestors. Any realistic concept of "species" in nature must somehow accommodate this inevitable heterogeneity across "gene trees." In the context of systematics, I have consistently championed the notion that reproductive isolation is (after all) a key to understanding why multiple gene trees tend to become shaped into coherent biological units.

A third broad conclusion from our molecular work is that speciation normally is an extended process rather than a point event in time. By comparing within- and between-species estimates of genetic divergence in mtDNA, we estimate that vertebrate populations often require about one to two million years to complete the speciation process, on average. Furthermore, this process is almost invariably associated with geographical separations.

Overall, I am now of the opinion that proponents of the PSC started an unnecessary brouhaha by arguing for a total overthrow of the BSC. In other words, the recent scientific literature on species concepts has falsely dichotomized reproductive and phylogenetic criteria as a basis for species recognition. From the genealogical insights initially prompted by mtDNA and interpreted in the light of coalescent theory, reproductive and genealogical underpinnings of biodiversity are, in truth, intimately related concepts, and mutually reinforcing.

Although intellectually stimulating, species concepts were mostly of academic interest until recently. That situation has changed dramatically with the heightened awareness that many elements of the world's biota are at immediate risk of extinction. Now there is an urgent need for means to identify and preserve distinctive biological units. Some of this concern already has been translated into conservation legislation. For example, under the U.S. Endangered Species Act, a "species" warranting lawful protection is any "distinct population segment" determined by wildlife experts to be in jeopardy of extinction. Due to the prominent social and economic (as well as biotic) ramifications of conservation efforts, questions of biological distinctiveness are no longer merely abstract.

Several genetic studies from my laboratory illustrate this point. In the late 1800s, a mammalogist traveling through the southeastern United States described a new species of rodent, the colonial pocket gopher *(Geomys colonus).* Apparently differing from the more common southeastern pocket gopher *(G. pinetis)* in details of pelage color and skull morphology, the colonial pocket gopher reportedly occurred only in Camden County, Georgia, near the coastal town of Saint Marys. The species' description was published in an obscure outlet, and for the next seventy years, the colonial pocket gopher went unnoticed and unstudied.

Then, in 1967, state biologists rediscovered a small population of gophers within the historic range of *colonus*. Numbering fewer than 100 animals in a single abandoned field, this population soon found itself on Georgia's endangered species list. Who could have guessed that these unassuming fossorial (digging) rodents, reclusive and nearly blind, were about to make a dramatic bid for the national limelight? This happened shortly thereafter when the navy announced plans to build a Trident nuclear submarine base near St. Marys. The gophers' existence clearly would be in jeopardy if the navy were permitted to go ahead with construction.

Knowing of our genetic work on southeastern pocket gophers, the Georgia Department of Natural Resources (DNR) requested that we quickly include *colonus* in our molecular assays. Happy to oblige, we compared samples of *colonus* to those of *pinetis* from throughout Georgia, Florida, and Alabama. In a wide variety of genetic assays, samples of *colonus* proved nearly indistinguishable from those of *pinetis* elsewhere in eastern Georgia and northern Florida. This was not due to poor resolution in the laboratory techniques, because each assay did uncover large genetic differences between regional populations of *pinetis* in the eastern versus western portions of that species' broader range.

The genetic findings were clear and their implications dramatic. Either the original description of *colonus* as a distinct species was in error or, in the seventy-year interim, *colonus* went extinct and was replaced by *pinetis* immigrants from adjoining areas. In science, one can never prove the null hypothesis, in this case that taxa do not differ genetically. Nonetheless, in the far more important, relational sense, extant *"colonus"* gophers appeared to be just another local population of *pinetis*. Still, to be sensitive to the animals' plight and to acknowledge the outside possibility that future genetic analyses might overturn these conclusions, we recommended that the DNR transplant the St. Marys gophers to a safer site, where they could continue to live in peace.

In a sense, our genetic findings in this case might be viewed as anticonservationist and prodevelopment. In effect, we had sided with the navy by removing a serious bio-legal obstacle to construction of the submarine base. However, the great strength of science as a means of knowing is its uncompromising objectivity. We reported the results as we found them. I've often joked that before releasing the findings, we should have extorted the navy! Our genetic work on the colonial pocket gopher had been done from the goodness of our patriotic hearts, without monetary support or remuneration. For the price of just one or two cruise missiles, I could have supported my laboratory's operations for an entire career!

Our multifaceted genetic analysis of this endangered species was published in 1982. To our knowledge, the paper was the first of its sort and quite ahead of its time (accounting, we prefer to think, for why it was initially rejected by some top-echelon journals!). Only later did molecular data become widely appreciated for the contributions they can make to taxonomic assessments of endangered species, including those embroiled in broader societal issues.

Our next involvement of this sort centered on the dusky seaside sparrow, another endangered species much in the public eye in the 1970s and 1980s. The dusky was a melanistic form of seaside sparrow described in 1872 as a distinct species *(Ammodramus nigrescens)*. Confined in native range to Cape Canaveral, Florida, its population had numbered in the thousands but then declined severely in the 1960s after much of its marsh habitat was altered for mosquito control or converted to pasturage. In 1966 the dusky was listed as endangered by the U.S. Fish and Wildlife Service, and it became a poster bird for the environmental movement.

In 1980 an intensive field survey found only six remaining birds, all males. In a last-ditch effort by wildlife personnel to preserve the dusky's genes, these

males were netted and placed in a captive breeding program with females of a related type of seaside sparrow, *A. maritimus*. The latter has a coastal range extending discontinuously from southern Maine to Texas, and the taxonomic status of its many populations is often controversial. For jurisdictional reasons, the particular females to which the dusky males were crossed came from the opposite side of the state, north of Tampa Bay. The intent of the wildlife managers was to produce hybrid females and then mate them back to the dusky males. In each successive generation of backcross progeny, the fraction of nuclear genes of dusky descent would increase predictably, and these dusky-like hybrids would then be released on Cape Canaveral to reestablish a wild population.

Due to the taxonomic uncertainties, we decided to examine genetic relationships in the seaside sparrow complex. Bill Nelson, who was to become my devoted research assistant for the next decade, spearheaded the effort. The last dusky seaside sparrow died in captivity in 1987, and its tissues were sent to us for genetic examination. In terms of mtDNA, this specimen proved to be closely similar to other seaside sparrows assayed from New York to South Carolina, yet it was quite divergent from populations along the Gulf of Mexico. In other words, there were two major matrilineal branches in the seaside sparrow's family tree, and these did not align properly with the taxonomic subdivisions upon which management programs had been based.

Under the BSC (or PSC), no arbitrary amount of genetic divergence between geographically separate populations can unambiguously define species status. Nonetheless, if the mtDNA phylogeny pointed reliably to overall genetic relationships in the seaside sparrow complex, then there was little empirical support for recognizing the dusky as a full taxonomic species (any more so than any other seaside sparrow population). Furthermore, even if dusky sparrows were deemed of *special* conservation interest, in the light of the genetic data a more enlightened captive breeding program would have crossed the remaining dusky males with nondusky females from Atlantic rather than Gulf Coast sites. That way, any birds reintroduced to Cape Canaveral would have been even more like the dusky in genetic composition.

Sadly, our findings meant little in pragmatic terms, because the last dusky sparrow had died, and the captive breeding program had failed. However, broader lessons emerged from the seaside sparrow experience. First, because taxonomic assignments shape our basic perceptions of the biological world, where possible they should be based on sound scientific information, particularly if societal or management issues are at stake. The dusky highlighted the fact that many species' epithets in modern use stem from cursory descriptions, often from the last century, of a few morphological traits of unknown genetic basis. Thus, an improved genetic understanding of biodiversity should contribute to better taxonomies and wiser conservation efforts. In the final analysis, biodiversity *is* genetic diversity, and molecular methods provide a powerful means to examine this variation directly.

A second lesson concerned the purview of genetics in conservation biology. In earlier conservation literature, if the topic of genetics was mentioned at all, invariably it was in the context of how best to preserve genetic variation in an endangered species, the rationale being that higher variation promotes survival over ecological or evolutionary time. This customary orientation stemmed from the heavy emphasis on captive breeding in the recovery plans for many endangered species. Small populations in zoos often suffer from inbreeding depression (poor fertility or viability of offspring) when animals have no choice but to mate with close relatives. Understandably, zoo biologists conveyed an implicit message in many of their writings: that "genetics" in conservation meant enhancing genetic variety via inbreeding avoidance.

Although concerns about inbreeding were certainly important, I saw several other underexploited opportunities for genetic perspectives in conservation biology, and not only for captive populations. Molecular genetic markers can reveal many conservation-relevant yet otherwise hidden facets about a species' natural history, behavior, phylogeography, and historical population demography in nature. The previous chapter described how genetic data on sea turtle migration, population structure, and evolutionary history informed management strategies for these endangered animals. In general, applications of genetic data abound in conservation biology, and for most of my career I have chanted that mantra to all who would listen.

However, while attempting to sell these novel perspectives, I also endeavored not to be a molecular chauvinist. Genetic insights, important though they are, invariably remain most helpful to any conservation cause when integrated fully with the vantages of ecology, ethology, demography, paleontology, and other traditional disciplines. Trained as an ecologist and natural historian, and remaining these at heart, I am keenly aware that the severe problems currently facing the Earth's biosphere will not be solved by mere genetic reappraisals of life, intellectually rewarding and helpful though such exercises can be. I tried to balance all of these perspectives in a 1996 volume, *Conservation Genetics,* that I coedited with colleague Jim Hamrick, summarizing about twenty of science's best available conservation-genetic case studies from nature.

Our genetic analyses of the colonial pocket gopher and dusky seaside sparrow challenged traditional species assignments underlying management plans for these endangered animals. However, in other cases our molecular analyses bolstered the otherwise shaky taxonomies of species on the edge of extinction. The previous chapter presented one such example, where mtDNA data reinforced a taxonomy, previously suspect, recognizing the Kemp's ridley as distinct from the olive ridley.

Another such example from our lab came years later, when DeEtte Walker addressed the phylogenetic distinctiveness of a threatened aquatic turtle, *Sternotherus depressus*. The word "depressus" refers to this pretty little animal's flattened shell, which gives it a squashed appearance. This musk turtle lives only in the headwaters of the Black Warrior River in northern Alabama, where its range is encircled by that of a common relative, the musk turtle, *S. minor*. Environmental threats to *depressus* include water pollution and hydrologic changes related to flood control and navigation projects. The otherwise secretive *depressus* gained a brief limelight in 1994, when it was showcased in *Life* magazine's montage of the 100 most photogenic of our nation's endangered species.

Stream beds in the Black Warrior system are characterized by large, horizontally stratified rocks that the turtles squeeze between for protection (a defensive strategy known as anachoresis). The animals' flattened shape is probably a recent adaptation for these special stream conditions. If so, the flattened shell might register only a few genetic changes from *minor,* or it might be an induced developmental response in hatchlings exposed to cramped crevices. Under either of these scenarios, the species status of *depressus* could be misleading if interpreted as an indicator of the population's overall genetic distinctiveness from related turtle species.

For her dissertation, DeEtte estimated a matrilineal phylogeny for all ten species of mud and musk turtles in the southeastern United States. To our surprise, *depressus* proved to be quite divergent from its nearest kin. Indeed, the genetic distance of *depressus* from its relatives typically exceeded geographic variation within the other named species. Thus, these little herpetological denizens of rocky stream hideaways in northern Alabama really are quite special genetically.

However, drawing firm management conclusions from genetic findings can be a more subjective business. One widely held assumption underlying phylogenetic appraisals of this sort is that distinctive populations or species warrant higher conservation priority than more humdrum ones, all else being equal. By this criterion, the flattened musk turtle and Kemp's ridley move up the priority scale. The basic rationale is that genetically divergent taxa contribute disproportionately to overall biotic diversity, and that maximizing the latter should be a primary conservation goal. But other considerations are important, too.

For example, species that make a large contribution to the health or stability of an ecosystem surely are of special conservation importance. These can range from top-echelon predators that crop prey species below levels where they overpopulate an environment, to obscure life forms that may remain inconspicuous until their disappearance unmasks their former ecological roles. Most importantly, enlightened societies should seek to preserve species simply because, like fine works of art, they enrich our lives. Conservation efforts for "charismatic megafuana" such as marine turtles, peregrine falcons, and cheetahs generate public support not so much because these species have unusual ecological impact but because they evoke wonder and are easy to love. But like connoisseurs of the arts, we must cultivate an appreciation for all manner of creatures, great and small.

Systematists have formally named about five million species, and many times that number (especially in the insect world) probably await discovery and description. In the coming decades, which species will we let live, and which will we sacrifice at the altar of "progress"? Shall the limited dollars earmarked for conservation be put toward saving the Florida panther, the Kemp's ridley turtle, or a patch of native Nebraskan prairie? The mere fact that such questions of conservation triage must be raised is a sad commentary on the degree to which a single species now exerts godlike authority over so many others. Like egoistic Romans pointing thumbs up or thumbs down to each plea from the coliseum floor, we diminish ourselves by forcing these spectacles of life and extinction.

15 Hybridization

Tobacco budworms—I had never heard of these homely little grubs, destined to become homely little moths. However, researchers at a U.S. Department of Agriculture (USDA) facility in Gainesville, Florida, were familiar with these agricultural pests. For years, they had hybridized two species in a crossing scheme designed to produce sterile offspring for release in large numbers to control natural budworm reproduction. In hybrids between *Heliothus virescens* and *H. subflexa,* males are invariably sterile, but females are fertile. In one long-term USDA experiment to test the basis of male sterility, female hybrids were mated to *virescens* males, their daughters were crossed back to *virescens* males, and this unidirectional backcross regime had been extended across ninety-one consecutive generations.

In 1981 this is precisely what I was looking for: a backcrossing scheme (in any taxon) in which females of hybrid ancestry had been crossed generation after generation to males of one of the parental species. I had scoured the literature and phoned around the world in search of this precise breeding design in any organism, only to find a perfect example practically in my own backyard. Why did I want later-generation hybrids from a bizarre set of matings that were unidirectional with respect to gender? It was to test, critically, the possibility of low-level "paternal leakage" of mtDNA.

In most animal species, each sperm cell carries a few dozen mtDNA molecules, and in some cases these can enter the mtDNA-rich egg during fertilization. However, no mtDNA from males had ever been uncovered in molecular assays of the resulting progeny. Maybe a father's mtDNA was present in his progeny but had escaped empirical detection due to the overwhelming preponderance of mother-derived mitochondria. However, if some "paternal leakage" of mtDNA did occur, this could have important evolutionary ramifications because it would open an avenue for genetic connections between otherwise separate mtDNA lineages within a species.

Backcross progeny permitted a refined test of the paternal leakage hypothesis. Think of it by analogy to a series of intravenous injections of a long-lived dye. The injected pigment might go unnoticed initially because it is diluted by a much larger volume of bright red blood. But after enough injections, the dye might finally become visible in the bloodstream. In our backcross analogue, the dye is sperm-derived mtDNA, each injection is a fertilization event, and the bloodstream is the extended maternal lineage across successive backcross generations. Even if paternal leakage is low on a per-generation basis, sperm-derived mtDNA might become enriched and eventually visible in the later-generation offspring.

We soon obtained budworms from Florida, but our molecular assays revealed only *subflexa*-type mtDNA in the ninety-one-generation backcross animals. Calculations showed that our assay conditions would have detected paternal leakage into these hybrids if as few as one mtDNA molecule in 25,000 per generation had come from the *virescens* sperm. At least for these species, results confirmed the conventional wisdom that for all practical purposes, mtDNA is maternally inherited. They also suggested that continued sterility of the hybrid males might be due to a "cytonuclear incompatibility," because the crossing regime had ensured that these backcross individuals were more than 99 percent pure *virescens* in nuclear genes, yet more than 99 percent pure *subflexa* with respect to mtDNA.

The point of this story is that hybrid organisms in general can be both fascinating in their own right *and* highly informative about evolutionary-genetic processes. The latter is true for several reasons. First, hybridizing populations or species are genetically divergent (by definition) such that many molecular markers normally can be uncovered for characterizing a hybrid gene pool. Second, because hybrids are a mixture of otherwise divergent lineages, their maternal and paternal genes might interact in magnified ways that make it easier to examine natural selection, genetic recombination, and other evolutionary processes. Third, a variety of sex-based asymmetries frequently arise in hybrid zones, and these intriguing gender biases can be dissected by studying matrilineal genes (like mtDNA) in conjunction with genes inherited biparentally on most nuclear chromosomes.

In searching the literature for hybrid settings to test for paternal leakage of mtDNA, I stumbled upon another example, one provided by Mother Nature herself. It involved topminnow fish (in the genus *Poeciliopsis*) in northwestern Mexico. There, hidden away in mountain arroyos, several species have been engaged (perhaps for thousands of years) in backcross hybridizations rather similar in effect to those forced upon the tobacco budworms by USDA researchers. The reproductive system of these tiny live-bearing fishes is complex, but it merits close inspection as a fascinating example of the outlandish evolutionary extremes to which organisms can go in their struggle for existence.

Like some piscine tribe of diminutive Amazons, there are absolutely no males in the species *Poeciliopsis monacha-lucida* (*Pml* for short). Instead, this all-female species reproduces by a "hemi-clonal" procedure known as hybridogenesis. In effect, *Pml* females are relentless "sexual parasites" on males of a closely related sexual species, *P. lucida*. These abused males provide sexual services to the *Pml* ladies but receive no lasting evolutionary remuneration for their efforts.

Each *Pml* female produces eggs via an atypical cellular mechanism that kicks out all of the nuclear chromosomes she had received from her father. Thus, although she carries her father's genes, she transmits none of them to what otherwise would be his grandchildren. Instead, she mates with a new *lucida* male, who likewise can become a father but not a genetic grandfather. Thus, each sexually exploited male contributes genes to his daughters' growth and development, but it is a dead-end contribution to the extended gene pool of the unisexual species.

I saw several research opportunities for these bizarre creatures. First, the possibility of low-level paternal leakage of mtDNA again could be tested critically. Rather like the budworm moths, the *Pml* females in effect had been engaged in a perpetual unidirectional backcross to males of a foreign species. Thus, if any *lucida* mtDNA had leaked into the system via sperm, it should be detectable in the *Pml* unisexuals.

Second, the phylogeny of any all-female species is nothing but its matriarchal genealogy, so the evolutionary histories of a *Pml* lineage and its mtDNA should in principle be one and the same. This contrasts with the situation in normal sexual organisms, where mtDNA's matrilineal record is only a tiny subset of the genetic legacy of a species. One ramification for the *Pml* fishes is that the evolutionary ages of the unisexual hemiclones might be estimated using empirical data from mtDNA. Conventional wisdom says that these lineages should be evolutionarily young, because a lack of sexual recombination presumably limits the adaptability of unisexual taxa.

Third, it was known that *Pml* had originated in evolution via hybridization between two sexual species (*P. monacha* and *P. lucida*). However, the direction and number of the original crosses were uncertain. By generating an mtDNA phylogeny for the group, it was likely we could discover how many hybridization events were involved and which of these two species was the female parent in the evolutionary origin of the unisexual topminnows.

I soon phoned Dr. Robert Vrijenhoek, a professor at Rutgers University who had devoted his research career to the natural history and genetics of *Poeciliopsis*. He was intrigued by the possibilities for mtDNA analysis, so we wrote and were awarded an NSF grant to address the issues raised above. With a graduate student, Joe Quattro, we soon headed off for the canyon lands of Sonora, Mexico. This was my first trip to the half-dozen isolated stream drainages that these reclusive topminnows call home, but it was the umpteenth trip for Bob, who knew this region like the back of his hand.

The stark hillsides and cliffs near Quaymas are spectacular where they meet the brilliant blue Sea of Cortez. With some trepidation, we pointed our rental jeep toward the interior mountains. This was prime marijuana-growing, drug-trafficking country, not particularly friendly to strangers. Another concern was whether topminnows could still be found. Their numbers had plummeted in recent years from predation by bass, tilapia, and other non-native fish introduced to these arroyos for food and sport.

Bob is another of the great characters it has been my privilege to know in science. An armchair philosopher, he also savors fieldwork, is a superb natural historian, and knows far more about the intimate lives of *Poeciliopsis* than you do about your closest friends. His favorite movie is *Up in Smoke* by Cheech and Chong, yet he is a sharp intellectual, equally comfortable delivering an advanced genetics seminar to an audience of stuffy academicians. If it suits his mood, Bob can be sophisticated or crude, laid back or compulsive, absent-minded or savvy, but seldom boring.

We spent several days rambling dusty roads through the hills, seining fish by day and camping by night. We met no bandits, but one minor incident epitomizes Bob and still makes me chuckle. We were driving through a small foothill town, minding our own business, when a uniformed policeman stepped in front of our jeep and commanded us to stop. He walked up to the driver-side door and demanded payment. Bob, dressed in a safari shirt and looking every bit the naive tourist, feigned ignorance of this gentleman's Spanish. Instead, with a wide smile and hearty handshake, Bob thanked the policeman in English for his concern about our welfare and then hit the accelerator. I reflexively ducked, but no bullets came flying, and we were soon out of sight.

Back in the States, over the next couple of years Joe genetically examined the topminnow collections. Several exciting findings emerged. First, as in the budworm moths, there was no evidence of paternal leakage of mtDNA. Second, the all-female *Pml* lineages had arisen on multiple evolutionary occasions from separate hybridization events between the sexual species. Third, we deduced that the successful matings underlying these *Pml* geneses always had involved *monacha* females with *lucida* males. Fourth, some of the unisexuals carried an extra chromosome set, and the genetic data revealed for the first time the detailed mechanistic route to this peculiar outcome. Finally, one unisexual topminnow lineage proved to be quite ancient, apparently having arisen from a hybridization event more than 100,000 generations ago.

Thanks largely to Bob's research efforts and those of his mentors and students, the peculiar unisexual topminnows have had a considerable impact on evolutionary thought, an influence vastly out of proportion to the fishes' modest numbers or drab appearance. Our genetic analyses were merely the latest of such studies. Tragically, these fish are now at risk of extinction because humans are altering their stream habitats and introducing exotic predators. What a shame it will be if these marvelous animals, uniquely instructive about broader evolutionary-genetic processes, are forced into extinction by the "wise primate" that has learned so much from their study.

I have returned to hybrid analyses frequently throughout my career, normally in animals with more conventional sexual orientations. Examples will illustrate some other kinds of secrets about nature that can be gleaned from molecular appraisals of hybrid creatures. One study involved *Lepomis* sunfish, about a dozen species of which inhabit eastern North America. These "panfish" range in size from adult dollar sunfish, no larger than a silver coin, to my old favorite, the bluegill, large individuals of which can indeed fill a small frying pan. When in breeding color during the spring and summer, sunfish are eye-catching. For example, pumpkinseeds display stunning orange and blue flecks across their flanks and brilliant streaks on their head and gill margins. Other species, such as the spotted sunfish, have a subtler beauty. Handsomely scaled in olive sprinkled with black spots, breeding adults seem dressed in a fine tweed suit.

Sunfish have a propensity to hybridize in various combinations, as first documented in the 1930s by Professor Carl Hubbs. Although these species normally retain their distinctive identities across broadly overlapping ranges, morphological oddballs do occasionally turn up. These atypical specimens show features generally intermediate to the pure parental species and are evidently hybrids. Hubbs discovered that first-generation (F1) hybrids of both sexes have low fertility. Nonetheless, these fish can occasionally produce later-generation hybrids as well, at least in artificial pond settings.

Whether the various sunfish species often exchange genes in nature remained an open question. If such introgression occurred, it most likely would involve F1 hybrids backcrossing to members of a pure species. Thus, these F1 hybrids could provide an evolutionary-genetic bridge for at least a trickle of migrant genes to move between species. I decided to test this possibility by examining natural sunfish populations in the Athens area. Five species occur regularly in our local waters—the bluegill, green, warmouth, redear, and redbreast.

For one entire summer, my daily research routine was to rise before dawn, head to a local stream or pond, and fish for an hour as the sun rose. In those days, each mtDNA workup necessitated a forty-eight-hour purification step through an ultra-centrifuge. We had two usable centrifuges, each with room for six tubes. This meant that on average we could process six fish per day, so that became my collecting goal. Upon visual inspection, most fish that I caught were easily assigned to one or another of the five sunfish species. However, eleven of the 1,000 captured fish were morphological misfits. I could guess which pair of species had produced each supposed hybrid, but until the genetic analyses were completed, I couldn't be sure, much less could I state definitively whether an individual was an F1 or a backcross specimen.

Each morning I clambered along the muddy, poison-ivy-cloaked banks of the Oconee River to find hidden fishing holes. The hybrid sunfish were rare and precious, like nuggets of research gold, and I could barely wait to genetically analyze those elite eleven. Actually, there should have been twelve. One morning as I rushed back to the lab with a prize hybrid specimen, I tripped on a vine while leaping a narrow ravine. Crashing to the stream bed below, I lay dazed for several minutes. Covered with mud and blood, I struggled to the car, thinking how ridiculous this academician must appear. Fortunately, none of the cuts and bruises was serious, so the only lasting pain was losing that treasured fish.

All of the species proved to be highly distinctive genetically, and the eleven oddities all proved to be F1 hybrids between various specifiable sunfish pairs. Thus, there was no genetic evidence of ongoing introgression between the species. Another interesting point was that in nearly all cases, it was the locally rarer of the two species in a hybrid cross that provided the female parent. Perhaps a paucity of local pairing partners or mating stimuli for females of rare species increased their likelihood of spawning with foreign males.

Over the years, several of our genetic analyses of hybrid populations have been sanctioned by wildlife managers. One study involved jumbo cousins of *Lepomis* sunfish, the *Micropterus* basses. The setting was a man-made reservoir on the Georgia–North Carolina state line. Lake Chatuge was originally inhabited by smallmouth bass and for many years supported an active sport fishery for this hard-fighting species. But in the late 1980s, something odd was revealed in annual surveys of the lake by Georgia's Department of Natural Resources. Smallmouth bass had nearly disappeared, replaced by spotted bass, normally found only in western portions of the state. Our efforts were enlisted to characterize the bass of Lake Chatuge from a genetic perspective and decipher what might have happened.

On an autumn evening in 1994, Bill Nelson and I accompanied about a dozen state fishery biologists to Lake Chatuge for a focused attack on the problem. Throughout the night, a small armada of electroshocking boats made forays to shoreline locations, periodically returning dockside to drop off the catch. Bill and I processed the incoming samples, labeling, taking measurements, dissecting organs, and storing tissues for genetic analysis. At the crack of dawn, we catalogued the last of 250 fish.

Later our genetic analyses of mtDNA and allozymes filled in several missing pieces to the Lake Chatuge puzzle. More than 99 percent of the bass proved to be spotted bass or their hybrids with smallmouths. Few pure smallmouth bass were discovered, although the Chatuge gene pool did retain a modest fraction of that species' genes, housed in various hybrid classes. Overall the genetic data confirmed that within only a few years, a dramatic shift in species composition had transpired in the lake, with spotted bass almost completely replacing the native smallmouth. This faunal turnover had been accompanied by hybridization between the two species in such a way that the original population of smallmouth bass had been "introgressively swamped" or "genetically assimilated" into what was now a predominantly spotted bass gene pool.

Where had the spotted bass come from? Georgia's state biologists suspect an unauthorized introduction, perhaps by sportsmen wanting to add spotted bass to the lake's fishery. If so, it worked, but only to the severe detriment of the native smallmouth.

There is an interesting genetic footnote to this story. In the last decade, Lake Chatuge has yielded several trophy-sized "spotted bass" that established state records for this species in North Carolina. Trophy bass and the tourist recognition they bring are taken seriously by all concerned. In December 1994 an eight-pound, fourteen-ounce bass was caught in Lake Chatuge, which would have eclipsed the prior state record for spotted bass by nearly three pounds. However, an official at the weigh-in station was suspicious. The fish looked a bit odd, and tissues were saved and later sent to us for genetic identification. They proved to have come from an F1 hybrid between a smallmouth father and a spotted bass mother. The upshot is that official certification was denied this specimen as a state record for the spotted bass. Although merely the purveyors of this genetic news, we were probably widely perceived as the bad guys.

Hylid frogs are another interesting group in the southeastern United States. Nearly twenty species in the region range from diminutive one-inch peepers, whose clear-toned songs are a welcome harbinger of spring, to five-inch Cuban treefrogs, whose calls resemble a grating snore. Variously streaked or spotted in motifs of gray, green, or brown, all true treefrogs (genus *Hyla*) have enlarged toe pads that enable them to climb adroitly. Males also have an inflatable vocal sac in their throat, which they use during the breeding season to issue a fine assortment of trills, whistles, honks, clinks, and clacks that fill the humid air on a warm summer's night. These mating calls are species specific, and herpetologists as well as female treefrogs can distinguish them readily.

Trip Lamb had such skills. When he joined my lab in the early 1980s, Trip was the prototypical graduate student ordained to succeed: bright and diligent, with great communication skills and deep insights into nature. Trip was a herpetologist since childhood, and he came to Athens with extensive knowledge on the identification and natural history of nearly every species of frog, toad, turtle, salamander, newt, gecko, skink, snake, and lizard to be found hopping, swimming, or slithering anywhere in North America.

Trip wanted a dissertation project that would give him experience in genetics *and* permit him to capitalize upon his "herp" background. Apart from his work on terrapins (described earlier), the project he settled upon involved two treefrog species common in wetlands of the southeastern coastal plain. The green treefrog *(Hyla cinerea)* has a lime green body with a white stripe down the side and a mating call resembling a clinking cowbell. The barking treefrog *(H. gratiosa)* has a rougher skin of dark green, often spotted, and a mating call resembling a dog's hollow bark. These species normally retain their separate identities, but according to the scientific literature, presumed hybrid animals of intermediate appearance and song had been common for more than twenty years in ponds near Auburn, Alabama.

Our interest in the Auburn site stemmed from a behavioral difference between the two species that we suspected might impact their population genetic structure. During the breeding season, *gratiosa* males typically issue mating calls while floating on the pond surface, whereas *cinerea* males sing from shoreline vegetation. In response to nighttime serenades by the all-male choirs, gravid females approach the ponds from the surrounding pinewoods and become amplexed by the ardent males. Amplexus is the treefrog's equivalent of coitus, wherein a male jumps on a female's back, clasps her tightly with his arms and thumbs in a wet embrace, and hangs on for dear life until she releases eggs, which he then fertilizes.

Given this breeding behavior, we wondered whether most hybridization events occurred in one direction only with respect to gender. There might be a sex-based asymmetry because, before female *gratiosa* can reach conspecific pairing partners in the center of the pond, they must hop a gauntlet of eager-to-mate *cinerea* males lurking along the shore. Female *cinerea,* on the other hand,

would be subject to few if any such untoward advances by *gratiosa* males floating out in the pond.

Sure enough, when Trip assayed more than 300 treefrogs from the Auburn site, almost all of the hybrids (which included many backcross and later-generation specimens) originally had stemmed from crosses between *cinerea* males and *gratiosa* females. How was this determined? Trip's genetic assays involved both mitochondrial and nuclear genes, which normally cleanly distinguish the two pure species. At the Auburn site, each hybrid animal was identified by virtue of possessing various mixtures of *cinerea* and *gratiosa* alleles at the nuclear genes, and the maternal parent was identified by the species-diagnostic mtDNA genotype that he or she had inherited from mom.

Trip's molecular study afforded a refined dissection of the genetic architecture of this hybrid population and was among the first to apply "cytonuclear" data to any hybrid setting. The analyses highlighted how molecular-genetic information can augment (indeed, outshine) field data in identifying particular classes of hybrid organisms, yet it also promoted the integration of morphological, behavioral, and genetic information. Even today, Trip's study remains one of the most lucid demonstrations of the power of cytonuclear analysis.

Our empirical analyses also prompted the development of a statistical and mathematical theory of "cytonuclear disequilibrium." Two of my colleagues at Georgia, Drs. Marjorie Asmussen and Jonathan Arnold, were trained as theoretical population geneticists, and we soon collaborated on a conceptual framework for interpreting nonrandom patterns of genetic association between nuclear and cytoplasmic genes. Marjorie, in particular, went on to extend our initial treatments greatly and establish herself as the leader in this large body of novel population-genetic theory.

One valid criticism of our work on natural hybrid zones over the years is that we had neglected to explore hybridization phenomena experimentally. Thus, when my next student, Kim Scribner, strode his 6' 5" frame into my lab, we decided to redress that criticism. I had known Kim for several years as the professional statistician at the Savannah River Ecology Laboratory (SREL). Also an experienced field biologist, Kim was ideally qualified to launch a series of technically demanding hybridization experiments.

For statistical and logistic reasons, the experimental design for Kim's dissertation required small organisms that could be reared in large numbers. The target species also must have a short lifespan, so that several successive generations could be monitored. Since we were partial to vertebrates, our options on suitable creatures were limited, but mosquitofish (*Gambusia holbrooki* and *G. affinis*) fit the bill. These inch-long fish thrive in captivity and have a generation length of about eighty days, and the females give live birth to about twenty-five to fifty young per pregnancy. These species also hybridize naturally in western Georgia, Alabama, and eastern Mississippi.

At the SREL site, Kim established separate *Gambusia* populations by introducing equal numbers of the two species into a replicated series of cement ponds and kiddy swimming pools. By periodically sampling several generations of mosquitofish over the ensuing two years, he monitored the progression of genetic changes in each population. Invariably, there was an initial flush of hybridization followed by a sharp decline in the frequencies of *affinis* genes. Thus, in these micro-evolutionary arenas, pitting two fish contestants against one another, the genes from *holbrooki* always won. In these ecologically simple environments, evolution was extremely rapid and the succession of outcomes surprisingly repeatable.

Not long thereafter, I received a telephone call that would lead to an interesting follow-up to Kim's experiments. On the phone was Dr. Steve O'Brien, a longtime friend and colleague, who was now inviting me to serve on a scientific advisory panel for the Biosphere 2 project, just outside of Tucson in Arizona's Sonoran Desert. I knew little about this facility beyond what I had read in the newspapers: Biosphere 2 is a futuristic glass and steel "greenhouse" closed off from material exchange with Biosphere 1 (the Earth and its atmosphere); the facility was envisioned as a scale-model experiment for how humans might someday colonize space; the venture also had been promoted as a grand experiment to explore "ecotechnological" frontiers for better husbandry of our planet; and much of the scientific rationale for the project had been under attack as inadequate or subsidiary to publicity-generating aspects of this privately funded mission.

Steve's telephone request was urgent. He had been asked by the directors of Biosphere 2 to convene a panel of accredited researchers to help design and implement critical scientific experiments within the enclosure. Ideally such research projects would capitalize upon unique ecological opportunities afforded by Biosphere 2 and perhaps blunt some of the media criticism of the project's operations. However, time was running short: Biosphere 2's first closure was scheduled to begin in three weeks. Eight Biospherians would enter the edifice, tighten the airlock doors, and live under sealed glass and public scrutiny for two years.

The panel's task sounded like great fun, so I immediately flew to Arizona to join about a dozen other population biologists to learn more about the project.

The Biosphere 2 facility in southern Arizona.

Located in beautiful desert surroundings, Biosphere 2 is something to behold. The graceful superstructure, enclosing three acres, consists of a triangular latticework of steel beams supporting large glass panels. It truly resembles some futuristic outpost on a distant planet. Inside, Biosphere 2 is literally and figuratively like a Garden of Eden, housing 4,000 species of animals and plants. There are marshlands, a miniature rain forest, desert, savanna, and stream, and an "ocean," complete with coral reef. The facility also includes living quarters, an agricultural area to feed the Biospherians, and a complicated "technosphere" of pumps, monitors, cooling systems, scrubbers, and other engineering wizardry designed to keep the environmental systems within boundaries suitable for life, all without benefit of physical contact with the outside world.

The tourist facilities surrounding Biosphere 2 are fascinating, too, if not bizarre. A visitor can order biomeburgers, habitat hot dogs, and planetary pizzas in the cafeteria. He or she can browse the gift shops for ecologically correct T-shirts and biospheric microcosms (miniature glass casings housing self-sustaining ecosystems). Thought-provoking films and guided tours explain relevant topics ranging from ecosystem functions to the technical design of space modules. Evocative sculptures, each named after a different Indian god, are scattered about the grounds. These were constructed of stainless steel salvaged from the Los Alamos atomic bomb project of the 1940s.

Our advisory panel was divided into small teams, each assigned to evaluate a different aspect of Biosphere 2's biota. My task was to assess the fish fauna. By reviewing the files, I discovered that most of the freshwater fishes introduced into Biosphere 2 were not from some natural community, but miscellaneous mollies, swordtails, and other tropical species purchased from a local pet shop! After compiling such information on the introduced flora and fauna, the panel wracked its brain to think of some scientifically useful set of experiments that might be set up quickly, before the facility's closure to the outside world. Alas, our lead time was too short, and little of substance was devised.

The one exception came about when I remembered Kim's research on mosquitofish. Perhaps we could repeat Kim's two-year-long hybridization experiments in the streams and marshes inside Biosphere 2. In these presumably more complex ecological communities, both species of *Gambusia* might persist indefinitely or perhaps display a different hybridization outcome than we had documented in the Spartan kiddy pools and cement ponds at SREL. I phoned Kim, and he quickly shipped *Gambusia* specimens for introduction into Biosphere 2. We returned two years later, just after the Biospherians completed their first mission, to collect the fish and assay them genetically.

As Kim and I entered Biosphere 2 with our nets and buckets, we hoped that our piscine charges had survived and reproduced. We also got a small taste of

The author netting mosquitofish from the freshwater marsh in Biosphere 2.

what the human Biospherians must have experienced for those two years. Dipping our nets repeatedly in the marsh and stream, we could feel the watchful gaze of tourists peering through the windows. I admit that we strutted a bit, pretending to be bona fide Biospherians rather than just a couple of visiting academicians. In any event, the mosquitofish were present in good numbers. Later our molecular assays revealed genetic outcomes remarkably like those found in Kim's surveys of the experimental populations back at SREL.

We hold no illusions that our hastily designed mosquitofish project inside Biosphere 2 represented great science, but the experience was enjoyable. Furthermore, far more important lessons were to emerge from the broader Biosphere mission. These have to do with how Biosphere 2 can foster a wiser appreciation of the ecosystem services in Biosphere 1. The following are excerpts and paraphrases from an editorial that I published in 1994 in the journal *Conservation Biology*.

The cost of the man-made technosphere that (marginally) regulated life-support systems for eight Biospherians over two years was about $150 million, or $9,000,000 per person per year. These services, such as atmospheric regulation and waste decomposition, are provided to the rest of us nearly cost-free by natural processes. If we were being charged, the total invoice for all Earthospherians would be an astronomical three quintillion dollars for the current generation alone! The sad irony is that human societies all too often have taken these ecosystem services for granted, acting as though we can befoul and overpopulate our planet without consequences.

During their two years of voluntary incarceration, the Biospherians became acutely aware of their dependence upon the fragile ecosystems within Biosphere 2. They would never have tolerated in their small enclosure the kinds of practices that are so widespread in Biosphere 1—massive deforestation, water and atmospheric pollution, the dumping of toxic chemicals, or overexploitation of renewable as well as irreplaceable resources. Nor would dramatic human population growth within Biosphere 2 have been tolerable, because precious commodities such as oxygen, food, and usable space already were stretched to the limit.

Exactly how many people Biosphere 1 can hold remains uncertain, but many signs indicate that we are quickly approaching reasonably sustainable limits. Ozone depletion and atmospheric pollution are global concerns, as are losses of nonrenewable resources such as usable groundwater reserves, soils, fossil fuels, and even biodiversity itself. Hunger, starvation, and conflicts over limited resources are recurring themes in many regions of the world. Yet, astonishingly, our species currently shows a net increase of more than 10,000 people every hour, a quarter million people each day. Within our children's lifetimes, the global population is projected to quadruple under current fertility rates. How much further the Earth's life support systems can be pushed remains to be seen, but all of us are unwitting guinea pigs in this reckless and utterly pointless experiment with global carrying capacity. Unlike the inhabitants of Biosphere 2, we have no outside source of rescue or escape. We can only save ourselves, hopefully through humane efforts at population control.

I concluded my editorial by writing what I perceived to be the real message from Biosphere 2. It may be fun and even inspirational to dream of colonizing other planets, but the harsh reality is that we have but one home, and it is getting untenably crowded. Whether based on ethical or purely utilitarian considerations, human societies must learn to value our Earth, and quickly. Like an astronaut's view from space, Biosphere 2 should give us a novel perspective and a renewed appreciation of Biosphere 1.

16 Alternative Reproductive Lifestyles

Natural historians like myself are endlessly fascinated by the planet's biodiversity and the multitudinous ways in which organisms go about their daily lives. However, not all behaviors of plants and animals in nature are evident to the naturalist's naked eye. Perhaps nowhere is this more true than in the often secret world of reproductive strategies, where field observations alone can provide poor or even misleading descriptions of nature's full panoply of procreative tactics. This is where molecular techniques can help. By identifying the true biological parents of particular offspring, genetic analyses of paternity and maternity can uncover many aspects of mating systems and reproductive behaviors that previously were hidden from view.

Vertebrate animals procreate in many ways, some of which seem downright outlandish by human standards. Earlier, I described our genetic work on a unisexual topminnow species that in effect has dispensed with males entirely via the evolution of a quasi-sexual reproductive mechanism. Over the years, my students and I likewise have used molecular markers to probe into the intimate lives and hidden affairs of a number of unusual animals with alternative reproductive lifestyles.

Behind every scientific paper is a saga of how an author came to recognize an unsolved problem, devised an empirical or conceptual approach toward its solution, and surmounted various hurdles of analysis. The staid format of the publication itself (Introduction, Methods, Results, and Discussion) implies an orderly, straightforward progression of thought and action, but seldom is this an accurate reflection of how the research truly was conducted. A personal account of most scientific studies would reveal a circuitous and often lengthy path to discovery, paved with happenstance, trials, and tribulations. A behind-the-scenes look at one publication on my résumé—"Evaluating kinship of newly-settled juveniles within social groups of the coral reef fish, *Anthias squamipinnis*"—will illustrate this point.

The sequence of events began in the fall of 1982, when I offered a two-week mini-course on molecular evolution at the University of Puerto Rico. In between lectures and laboratory exercises with biology students on the Río Piedras campus, I had time to tour the island. One trip took me to a marine station on the southwest coast, where I met Doug Shapiro, a fish behaviorist, who in a former career had been a practicing physician.

Doug's research centered on social behaviors in group-living species, with special focus on marine fish that can change sex during their lifetimes. Although it may seem bizarre to us, gender switching is standard procedure for many fish in the sea. Some species are protogynous hermaphrodites, meaning that an individual begins life as a female but later may convert to a functional male. Other species are protandrous hermaphrodites, in which an early-life male may later become a female. A few species are synchronous hermaphrodites, wherein individuals are part male and part female at the same time, producing functional sperm and eggs simultaneously. Behavioral ecologists are interested in how an animal's lifelong reproductive success might be related to its social and physical environment, and hermaphrodites are particularly interesting in this regard.

Doug studied a small sea bass, *Anthias squamipinnis,* abundant on Indo-Pacific coral reefs. These red fish typically hover above a reef in discrete schools of about ninety adults each, females outnumbering males eight to one. When a male dies, within a few days a female begins to transform into a replacement male. From experiments conducted at his study site in the Red Sea, Doug had found that the pace and pattern of transformation are also functions of group composition and of interactions between females at various levels in the group's social hierarchy.

In *squamipinnis,* as in many other marine fish, spawning produces pelagic larvae that drift in the open sea for days or weeks. Conventional wisdom is that progeny from separate spawns must mix thoroughly during the pelagic phase, such that each larval cohort finally "settling" on a reef represents a more-or-less random genetic draw from the local gene pool. Doug had come to question this view. He suspected that *squamipinnis* larvae from a single pair of spawning adults might often stay together through the pelagic phase and settle jointly, so that each school of fry would consist primarily of full sibs. If so, ecologists would have to consider kin selection (natural selection acting on relatives) as another factor potentially directing behavioral evolution in this and other group-living marine fishes.

Doug wondered whether his hypothesis might be critically tested using allozyme techniques. I thought so, provided that we could identify variable genetic markers to assess kinship in this species. We would need to collect several juvenile schools immediately following their settlement, and we would also need to modify existing lab methods so that fry the size of a pinhead could be assayed individually. Doug had a few *squamipinnis* in the freezer, so I took these back with me to Georgia to develop the laboratory assays. The species proved highly variable genetically, and after some trial and error I was able to adapt our protein-electrophoretic procedures to accommodate tiny bits of tissue. Based on these preliminary results, Doug and I submitted a grant proposal, soon funded by the National Geographic Society, to support our collecting efforts in Israel.

As luck would have it, at about that same time I was invited to Israel to participate in a two-week conference on molecular evolution. Professor Eviatar Nevo, the organizer, promised a nice mix of meetings and educational tours. The symposium was scheduled for March, which happened to coincide with spawning and settlement of *squamipinnis* in the Red Sea. Accordingly, Doug and I agreed to rendezvous in southern Israel following the conference to collect fish for our project.

A unique blend of cultural and ecological experiences made the Israeli symposium highly enjoyable. Work sessions (each in a different city—Jerusalem, Haifa, Tel Aviv, Beersheba) were interspersed with sightseeing excursions. We visited the Old City of Jerusalem, the Dead Sea, the Mount Carmel caves, and a kibbutz on the Sea of Galilee. My favorite stop was at Masada, a spectacular butte soaring above the Negev Desert. We took a cable car to the top, but in ancient times the only access was via a steep trail. Two thousand years ago, in a mountaintop fortress, 1,000 Jewish Zealots opposed to the Roman domination of Palestine held off a 10,000-man siege for three years. When Masada fell, the last defenders committed suicide as the Roman legions rushed across a huge stone rampart they had constructed from an adjoining mountain.

Dr. Nevo clearly relished this opportunity to show off his country and expound on his own research ideas. Eviatar is known for his thesis that natural selection accounts for even the finest patterns of geographic variation in biotic features. Seeing Israel firsthand made this perspective easier to appreciate. Packed within this tiny realm are ecological settings ranging from arid plains to verdant valleys, from blistering sea-level deserts to chilling mountains. It is as if many of the diverse geological and ecological features of the continental United States were compressed into a state the size of New Jersey. Superimposed on this ecological mosaic are plant and animal communities distinctively adapted to these varied settings.

After the conference, I took a bus south to Elat, Israel's port on a northern arm of the Red Sea. My plan was to meet Doug, who was traveling from Puerto Rico. I arrived at the Steinetz Marine Lab on the appointed date only to learn that Doug was delayed indefinitely because our baggage and dive gear had been lost en route.

While passing time in Elat, I couldn't help noticing that many people were carrying binoculars. They proved to be birdwatchers from around the world, beckoned to Elat like Muslims to Mecca. Elat in late March is one of the premiere birding hotspots on Earth, because avian migrants from across Africa return to their summer homes in the northern hemisphere. These birds funnel up the narrow Gulf of Aqaba to hit Elat in great waves that attract amateur and professional ornithologists alike. The fact that many of the species nest in relatively inaccessible Middle Eastern countries such as Iran and Iraq only makes Elat that much more alluring to Western ornithologists.

I hooked up with some knowledgeable birders, and each day we scoured the local environs for the latest avian arrivals. We scanned the beach, hiked nearby mountains, and monitored the desert. Professional teams of bird-banders from Sweden, England, and Germany were camped in an agricultural valley along the Jordanian border. They let us patrol their lines of mist-nets and inspect each ensnared bird before it was banded and released.

One warm afternoon, a black streak appeared high in the sky, from horizon to horizon. Binoculars revealed that it was composed of hundreds of birds, each a migrating hawk or eagle. These raptors avoid flying over large bodies of open water such as the Mediterranean Sea and, thus, tend to accumulate along shoreline corridors. The configuration of the Middle East is such that migrating raptors from across Africa are channeled through constriction points like Elat before splaying out to reach their nesting sites in Europe and Asia. Perhaps nowhere else in the world can so many hawks and eagles, belonging to about twenty species, be seen at one time.

After a week's delay, Doug finally arrived in Elat, but only after I had been forced by other commitments to return home. He collected several schools of juvenile fish and shipped them to Georgia. Several months later I completed the lab work and analyzed the genetic data. It turned out that each school of *squamipinnis* consisted of unrelated fry from multiple spawns. Doug was disappointed by this outcome: a finding that each school was composed of full-sibs would have been more revolutionary. I was delighted simply that we had managed to pose and then answer a tough question about nature. We crafted a manuscript and steered various revisions through the external review process.

Finally accepted, the paper was published a year later in the journal *Evolution*. Such is the inside story of the kind of effort underlying a typical study in the field of population genetics.

Some research projects involving odd means of animal reproduction have fallen into my lap. Such was the case in 1993, when Drs. Jim Loughry and Colleen McDonough of Valdosta State University wrote to ask whether I would collaborate in their ongoing studies of nine-banded armadillos. These mammals, covered in leathery armor, are native to Central and South America but in the last two centuries have expanded their range into the southeastern United States. For years, Jim and Colleen had been studying the ecology and behavior of armadillos near the Tall Timbers Research Station (where I had obtained tower-killed birds years earlier).

Even stranger than the armadillo's appearance is its reproductive system, nearly unique in the vertebrate world. Each female produces a litter of genetically identical pups, usually quadruplets. These clonal babies trace to a single fertilized egg that divided into four sets of daughter cells before embryonic development was initiated in their mother's womb. The first hint that armadillos reproduce by this "polyembryonic" mode came early in the twentieth century, when it was noticed that each litter consists of same-sex progeny. Other organisms (humans included) occasionally produce monozygotic twins, but armadillos are the only mammals thought to produce clonemate litters as standard procedure. Polyembryony differs from other forms of clonal reproduction (such as parthenogenesis) in that the members of a clone are intragenerational rather than intergenerational. Thus, progeny within a litter are genetically alike

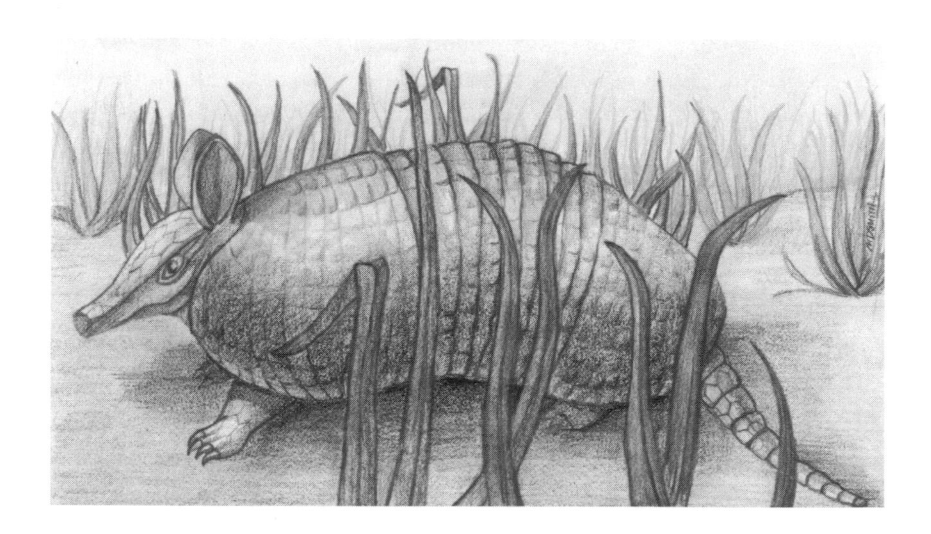

but differ in genetic makeup from their mother and father, just as do offspring in normal sexual species.

To an evolutionary biologist, obligate polyembryony is intriguing because the adaptive significance (if any) of the phenomenon is obscure. As an ecological tactic, polyembryony lacks the presumed advantages typically associated with either sexual or asexual reproduction. Sexual reproduction creates genetically variable offspring, some of whom might be adapted to local environmental conditions, but polyembryony produces littermates with only one genotype. Asexual reproduction replicates a mother's genotype in progeny, but polyembryony produces multiple copies of a previously untested genotype. In life's reproductive raffle, it is as if polyembryonic animals have adopted a boneheaded strategy of purchasing multiple lottery tickets of the same number.

The paradox of polyembryony has several potential solutions. One intriguing possibility is suggested by the fact that polyembryony is common in parasitic wasps. In some of these species, a female oviposits a single egg into a host species' egg, which later develops into a caterpillar, an abundant food source for the parasite's developing young. In effect, each mother wasp reproductively colonizes a tiny patch of favorable habitat (a host egg), yet is able to leave many kids by virtue of the subsequent polyembryonic cellular divisions in a richer ecological setting (the caterpillar). For both the parasitic mother and her clonal progeny, the strategy of polyembryony may simply be making the best of the available reproductive circumstances.

Another evolutionary explanation for polyembryony invokes the idea of nepotism— favoritism toward kin. In the case of armadillos, perhaps the quadruplets remain together for an extended time to assist one another, for example, in cooperative foraging, denning, or detection of predators. If so, polyembryony might have evolved under kin selection as an adaptive strategy, provided that the average fitness benefits accruing to the clonemates outweighed those of alternative reproductivc tactics. Jim and Colleen had been searching for armadillo behaviors consistent with nepotism, but without success. Their efforts were hindered by the fact that clonemates can't be unambiguously identified by external appearance or from field observations alone. This is where we came in.

In the 1980s a variety of "DNA fingerprinting" methods had been introduced. These laboratory assays reveal "hypervariable" DNA profiles that uniquely earmark each individual in a sexually reproducing species, much as do conventional human fingerprints. The assays again involve an electrophoretic step, in this case of DNA isolated from cellular nuclei. The hypervariable gene regions are displayed as complex gel bands, which resemble the familiar bar codes identifying shopping items. DNA fingerprints had become widely used in human forensics, for example, in courtroom trials where the source of a blood or semen sample is at issue. Population geneticists likewise were beginning to explore the possibilities for DNA fingerprinting in wildlife studies.

For the armadillos, it was likely that DNA fingerprinting techniques would enable us to identify clonal littermates in the Tall Timbers population. The person I recruited to the task was Paulo Prodöhl, an affable Brazilian with a Ph.D. degree from Queen's University in Belfast, Northern Ireland. Within a year, he developed a battery of DNA-level markers and assayed more than 300 armadillo tissue samples (tiny ear clips). Paulo's approach involved the assay of microsatellite loci, a class of genes, each typically occurring in many different forms in a population such that genetically unique animals can be discriminated with extreme precision.

Paulo thereby identified armadillo clonemates from Tall Timbers and mapped their spatial distributions at the time of capture. The results were consistent with Jim and Colleen's suspicions that armadillo siblings generally do not remain near enough to one another to promote cooperative behaviors that might lead to nepotism and kin selection. Paulo had solved the problem of identifying clonal sibships in nature, and his data helped to eliminate a competing hypothesis for the evolution of polyembryony in armadillos. However, the broader evolutionary paradox remained.

One clue to the polyembryony puzzle in armadillos may come from the parasitic wasps described earlier. In female armadillos, the uterus is oddly constricted. After a successful mating event, a single blastocyst (pre-embryo) remains quiescent for several months at the tiny uterine implantation site, only later dividing to produce polyembryonic fetuses. We now suspect that polyembryony evolved in armadillos as a fitness-enhancing means of increasing litter size by circumventing a structural limitation imposed by a uterus with only one implantation site. Thus, in both the wasps and armadillos, polyembryony may have arisen as a selectively advantageous mechanism to sidestep a peculiar reproductive bottleneck. This idea, though speculative, illustrates that intriguing hypotheses in comparative biology sometimes can arise from the most unlikely of evolutionary juxtapositions.

However, even if the uterine-constraint hypothesis is correct, it offers only a near-term explanation for polyembryony in armadillos. A deeper question would remain: why did these creatures evolve this unusual uterine configuration to begin with? This is another appealing element to me of the evolutionary sciences—they often raise (and explicitly distinguish) explanatory hypotheses involving various levels of proximate and ultimate causation.

Apart from identifying particular individuals and clonemates in nature, by the late 1980s microsatellite assays and related DNA-fingerprinting approaches also were starting to revolutionize the study of animal mating systems. By properly interpreting nature's bar codes in the context of basic hereditary principles, researchers could use these genetic data to deduce biological parentage: for example, who did (or did not) sire particular offspring? An accumulation of such data can reveal the genetic mating system of a population and lead to insights about reproductive strategies that might not have been evident from field observations alone.

Consider, for example, studies on the perching birds (Passeriformes). From field observations, most species (90 percent) had been thought to be socially monogamous, such that each faithfully mated pair supposedly produced a tight nuclear family of full-sib progeny. However, numerous molecular analyses over the past twenty years have overturned this conventional wisdom. Many passeriform species have proved to be genetically polygynous: males often mate successfully with multiple females. The evidence came from genetic paternity exclusions, when particular males had not necessarily sired all offspring in their respective nests. The remaining progeny must have come instead from instances of cuckoldry or "extra-pair fertilization" (EPF), often by other nesting males. Also documented by molecular markers were cases of egg-dumping, or "intraspecific brood parasitism" (IBP), wherein a female occasionally lays eggs in another's nest so that the resulting offspring are reared by foster parents. In general, genetic studies on numerous avian species, conducted by many laboratories especially during the 1990s, began to clarify a fundamental distinction between the apparent "social mating system" of a population and the true "genetic mating system," which often may include surreptitious mating events leading to progeny production.

By the late 1990s, the molecular literature on avian (and mammalian) mating systems was quite large. Thus it was surprising that almost no such studies had been conducted on the amphibians, fishes, or reptiles, despite the fact that these "cold-blooded" vertebrates collectively have a vast array of reproductive behaviors simply begging for exploration by the genetic tools of this new breed of natural historians. Research frontiers that are both spacious and technically amenable to exploration are hard to identify because they quickly tend to be colonized by entrepreneurial scientists. Yet in 1995, when I first thought about it, here was a rich and untapped research arena, and better still, it was situated squarely at that wonderful research juncture between the hard genetic sciences and good, old-fashioned natural history. By studying the genetic mating systems of fishes and other cold-blooded vertebrates, we might learn much about the intimate reproductive lives of some exceptionally fascinating species.

There were additional appeals to doing genetic work on mating systems in cold-blooded vertebrates. For example, most fishes are highly fecund, with clutches often composed of hundreds or even thousands of young. Such large broods raise challenging technical issues regarding optimal sampling design and statistical analysis of genetic data that simply had not arisen in avian and mammalian studies. Furthermore, in creatures with large family sizes, several molecular-level issues also could be tackled more readily, such as detailed assessments of mutation rates and patterns.

Our first excursion into this realm was spearheaded by Adam Jones, a shy but brilliant graduate student, who together with Paulo was instrumental in converting my lab to the newer microsatellite techniques. Adam's dissertation research would involve assessing genetic mating systems in pipefishes and seahorses (family Syngnathidae), an adorable group of marine fish in which it's the males (rather than the females) who get pregnant.

The phenomenon of male pregnancy, otherwise rare in the biological world, is the universal state of affairs in the approximately 250 living species of Syngnathidae. During each mating event, a female deposits her eggs into a specialized brood pouch on her mate's underside. The male then assumes all parental duties. He fertilizes the eggs, provides his developing brood with nutrients, aeration, and osmoregulation, and protects the young until they are liveborn weeks later. The male's brood pouch in most species is fully enclosed, so a care-giving male has complete assurance of genetic paternity for his kids, probably a cardinal element in the evolution of this high-investment male strategy in reproduction.

This paternity assurance for a pregnant male was also a key factor in our genetic analyses of syngnathid mating systems. By knowing a brood's father, we could deduce the maternal genetic contribution to each embryo by subtraction. In other words, we could figure out who the mother was of each embyro within each male's brood pouch. Genetic maternity analyses are a reverse counterpart to the paternity analyses routinely conducted in human forensic labs to deduce the biological father of a woman's child.

Each pregnant pipefish or seahorse carries scores or hundreds of embryos. Through genetic analysis, we wished to learn how many females contributed to a male's brood, and also whether a given mother might have dispersed her clutch of eggs by successfully mating with two or more males in close succession. From an accumulation of such genetic information for each population studied, we could distinguish among three candidate mating-system outcomes: monogamy, wherein each brood normally results from a faithful male-female pair; polyandry, wherein a female often mates with two or more males, but a male typically breeds with only one female; and polygynandry, in which both sexes often mate successfully with multiple partners during a specified breeding period.

In "sex-role-reversed" species like pipefishes and seahorses, this monogamy → polyandry continuum of possibilities is a mirror image of the monogamy → polygyny continuum typically observed in mammalian and other species with "normal" sex roles, wherein maternity is assured and female reproductive investment is high. Genetic polygyny is defined as the successful mating of particular males with multiple females, and of each female with one male. Evolutionary biologists ever since Darwin have been keenly interested in mating systems because in theory they relate to a host of biologically interesting features: the relative investment in offspring by male and female parents, reproductive rates of the two genders, the relative intensity of sexual selection on males and females, and the degree of sexual dimorphism within a species.

Traditional theory predicts, for example, an association between parental investment strategies and the evolved behavioral tendencies that underlie a realized mating system. In mammals with "conventional" gender roles, a female's investiture in progeny is high if for no other reason than her pregnancy. When males expend heavily in offspring as well (think of foxes, where both parents bring food to pups), monogamous behavioral tendencies may be the evolved norm. However, when female reproductive investment greatly exceeds that of males (think of nursing elk), polygyny is more likely. In such species, males tend to be eager maters because their reproductive contribution ends with copulation, whereas females tend to be coy (choosier) because they alone will bear the burden of pregnancy and offspring care. In a nutshell, males have been selected over the eons for inherent sexual impudence because a male's genetic contribution to the next generation is strongly correlated with how many mates he obtains. Females, by contrast, have been selected for relative sexual reticence because their genetic fitness is influenced by mate quality more so than quantity. This traditional hypothesis of asymmetry in genetic fitness as a function of male-versus-female mating strategies is known as Bateman's gradient.

In the topsy-turvy reproductive world of pipefish and seahorses, male investment in offspring production in some cases actually exceeds that of females, meaning that if either gender has evolved a behavioral disposition to be more discriminating in mate choice, it is likely to be the male. Thus, any departure from monogamy (or polygynandry) is likely to be in the direction of polyandry rather than polygyny.

Such logic extends to theoretical expectations regarding the intensity of sexual selection and the degree of sexual dimorphism. In polyandrous species, sexual selection stemming from greater male choosiness and/or from stronger mate competition among females may lead to the evolution of female "secondary sexual traits" such as mate-attractive body ornaments or fighting ability. Indeed, in pipefish species with sexual dimorphism, almost invariably it is the female who is more adorned. This reverses the familiar situation in polygynous mammals and birds, where any secondary sexual features typically are elaborated in males (think of a bull elk's rack, or a peacock's showy tail).

For his dissertation, Adam explored these and related topics in several pipefish and seahorse species from North America, Europe, and Australia. For each surveyed species, he developed and employed microsatellite markers to deduce the genetic parentage of particular broods. He mapped the arrangement of embryos from different females within each male's pouch, tallied the number of deduced mothers who contributed to each brood, and examined whether particular females had mated with more than one male. From an integration of genetic and natural history information, he thereby deduced the local genetic mating system of each species.

Consistent with the predictions of mating-system theory, Adam discovered that sexual dimorphism was greater in genetically polyandrous than in monogamous or polygynandrous species. Furthermore, he showed in laboratory mating experiments that Bateman's gradient applied in mirror-image fashion (exactly as predicted by theory) for these sex-role-reversed species. Thus the genetic fitness of females but not of males was strongly correlated with their numbers of mates.

In these and other fish species, Adam also took advantage of the large clutch sizes to illuminate several molecular-level phenomena. For example, he discovered sex-biased patterns of genetic recombination between linked genes in one species. In another, he documented the molecular basis of a microsatellite null allele. In a third species, he characterized the nature and frequency of "clustered mutations." These are new variants arising premeiotically in the germ-line of a parent, and, hence, they can appear in multiple embryos within a brood. Such genetic dissections also illustrated a broader point—when using DNA or protein markers to address natural-history questions, one must remain cognizant of the molecular-level mechanisms underlying the genetic polymorphisms utilized. Indeed, it is often the interplay between mechanistic and population-level understanding that makes evolutionary-genetic studies so intellectually challenging and exciting.

By assuming the arduous duty of pregnancy, pipefish and seahorse males have gone to extreme lengths in offspring care. However, dedicated parents can be found in many other fish species as well. Some fish are oral brooders, routinely carrying eggs and young in their mouths for protection. In other fish species, parental care (usually by males) takes the form of guarding, cleaning, and fanning the fertilized eggs in a nest constructed in aquatic vegetation or in the substrate of a stream or pond. In effect, these males have external pregnancy. In North American sunfishes, for example, a male scoops out a shallow, platelike depression in mud or sand, attracts one or more gravid females to the nest, fertilizes the eggs she lays, and then tends the resulting larvae for about a fortnight before the fry disperse.

In fish with external fertilization, a new wrinkle is added to parentage assessments because a primary nest-tender might sometimes be cuckolded by other males. Such stolen fertilizations could occur, for example, when one or more "satellite" or "sneaker" males swim onto a nest during a spawning event and release sperm that fertilize some fraction of a female's newly released eggs. Or the fertilizations might have been stolen from the primary nest-holding male by another attendant male nesting nearby. Such possibilities are not merely conjectural. Field observations suggest that "reproductive parasitism" is a routine part of the natural histories of many fish species.

Prior to our own molecular work, one of the few genetic studies on fish parentage helped to document high frequencies of fertilization thievery in colonies of bluegill sunfish in Canada. There, field observations had shown that male bluegill come in two distinct reproductive forms: parental males, who first mature at about seven years of age and compete for nesting sites within a colony; and cuckolder males, who mature precociously at age two and then may steal fertilizations in the resident males' nests. When small, these potential cuckolders simply dart into a nest and release sperm as the female extrudes her eggs. When older and larger, cuckolders gain access to a nest by mimicking females in coloration and behavior. In one fish colony in Ontario, genetic analyses revealed that more than 50 percent of bluegill fry had been sired by reproductive parasites—the sexually precocious juveniles, as well as these adult female impersonators!

External fertilization in conjunction with male nesting behavior also opens windows of opportunity for several other bizarre reproductive tactics in fishes. Over the last several years in my laboratory, a succession of excellent students (in addition to Adam Jones) and postdocs (most notably a delightful Texan, Andrew DeWoody) have employed molecular markers to unveil local genetic mating systems in numerous fish species with extended parental care of young, as well as in some other vertebrate groups such as turtles. Apart from genetically documenting several reproductive behaviors that were poorly known previously, the assays also have permitted quantified estimates of the fitness consequences of alternative mating strategies. Among the many interesting reproductive phenomena revealed in our molecular assays are the following:

Multiple mating by males. In several species of sunfish, our genetic analyses revealed that two or more (up to a dozen) different females often have spawned in a male's nest. Thus, a typical sunfish nest includes a mixture of full-sib and half-sib progeny. We have documented similar kinds of within-nest family structure stemming from multiple maternity in species of sticklebacks, darters, and others. For example, exceptional reproductive promiscuity was displayed by the sand goby, a tiny marine fish in which adult males tend nests positioned under a stone or mussel's shell. From the genetic evidence, most of the nests contained embryos from three or more mothers, and many of the nest-tending males also had been cuckolded (see below). These sexually busy fish apparently make the most of their brief time on earth (one year).

Multiple mating by females. In some species of nest-tending fish, we were also able to document with genetic markers that particular females had spawned successfully in more than one nest. Due to the logistical constraints of sampling all relevant nests in which a female may have laid her eggs, the instances that we uncovered probably represent only the tip of the iceberg of this phenomenon.

Cuckoldry by surreptitious males. In most of our surveyed species, at least some of the attendant males proved not to have sired all of the young in their respective nests. Instead, a significant fraction (often 10 percent or more) of the resident offspring were the result of cuckoldry, either by neighboring nest-tenders or by sneaker or satellite males. As mentioned, sand gobies in particular had unusually high rates of genetic cuckoldry.

Egg thievery by nest-tending males. In sticklebacks, males use a gluelike substance secreted from their kidneys to construct hollow, vegetated nests into which females lay discrete clusters of eggs. Surprisingly, our genetic data for the fifteenspine stickleback also pointed to the likelihood that an occasional egg cluster had been stolen by the resident male from another male's nest.

At first thought, egg thievery would seem to be an ill-conceived (pun intended) fitness strategy for the larcenist because in effect he is thereby adopting the offspring of a rival male. Nonetheless, in earlier field studies, males had been observed carrying stolen eggs in their mouths, and ad hoc explanations were advanced for the significance of this behavior. Perhaps the pilferer is stocking his own larder, because nesting males often eat some of the local embryos and fry. Perhaps the stolen eggs are those that the thief himself had fertilized in an earlier cuckoldry raid, so the male would be raising his own kids after all. Another possibility, which we favor, comes from the suspicion that females of many fish species prefer to spawn in egg-containing nests. Indeed, in some darter species, a breeding male's fins or body are adorned with egg-mimicking spots that presumably help elicit a spawning response from passing females. Thus perhaps egg larceny in male sticklebacks is an evolved behavioral tendency selectively advantageous to the thief because it "primes the pump" of female egg-laying in his nest.

Nest takeovers. In rare cases, a resident male proved by genetic analysis to have sired none of the embryos or fry in his nest. Such cases probably evidence nest takeovers by a foreign male after the former resident departed or, perhaps, was evicted. Nest eviction might be especially common in species or in ecological settings where suitable nesting sites are in short supply.

Cuckoldry by females. In the largemouth bass, which proved to be mostly genetically monogamous, some nests nonetheless contained a few offspring that were not the progeny of the guardian female. This species is also unusual in that both sexes tend the fry. Thus the "aberrant" offspring can be interpreted as resulting from "female cuckoldry," wherein a second female appears to have laid a few eggs in the nest and thereby usurped for her own children some of the guardian services of the foster mother.

Filial cannibalism. Many fish have long been thought to eat their own young on occasion, a phenomenon known as filial cannibalism. However, given what we now know about the reasonably high frequencies of nonparentage by juvenile-tending adults (via cuckoldry, egg thievery, and nest takeovers), an intriguing possibility remained that perhaps most fish are not eating their own biological young after all, but instead preferentially consume those of other parents. To test this hypothesis genetically, we examined recently consumed fry within the guts of adult, nest-tending tessellated darters. The data genetically confirmed, for the first time, that filial cannibalism does indeed occur.

Fish divorces. Unlike the passeriform birds, few fish species are socially monogamous, but among those that are, does genetic monogamy hold as well? Our genetic studies of the Western Australian seahorse showed that the study population is indeed genetically monogamous within a breeding period, but also that a high proportion of the couples (more than 40 percent) get "divorces" and "remarry" between successive reproductive bouts. Largemouth bass likewise often proved to have remarried after losing a spouse.

Multiyear sperm storage by females. Using genetic markers, we have uncovered interesting reproductive phenomena in some other cold-blooded vertebrates as well. For example, my student Devon Pearse collaborated with Dr. Fred Janzen at Iowa State University to examine reproductive patterns in painted turtles. Genetic parentage analyses of successive clutches of marked individuals documented that a female can store and utilize sperm for at least two years following a mating event. This remarkable capacity for long-term storage of viable sperm in the reproductive tract of female turtles was known from earlier studies of isolated animals (who may continue to produce young in captivity), but the phenomenon had not been previously documented in nature.

Most of our genetic parentage studies of the cold-blooded vertebrates entailed formidable and often ingenious collecting efforts. For example, redbreast sunfish males on the nest are skittish and cannot be approached during daylight. Our method was to spot a nest from the bank, place marker stakes nearby, and return at night to capture the sleeping attendant by using a backpack electroshocking rig. Each nest was then scooped out and returned to the lab. There hundreds of fish embryos and larvae were identified under a dissecting microscope and painstakingly sorted from the mass of debris and vegetation. Only then could the equally arduous genetic assays begin.

To me, these kinds of natural history studies, combining high-tech molecular assays, demanding fieldwork, and refined procedures of genetic data analysis, are gratifying. They yield answers to previously intractable questions about animal reproductive behaviors in nature. Furthermore, the exercises themselves offer sharp intellectual challenges. Working through the intricate logic of parentage and mating-system analysis via genetic markers is not unlike solving a complicated jigsaw puzzle. When all the pieces finally are made to fit properly, what emerges can be a beautifully clear and sometimes unexpected picture about natural-history processes. Finally, such findings often contribute to broader conceptual understandings of the ecology and evolution of animal mating systems and alternative reproductive tactics.

Several of our publications on mating strategies in pipefish, seahorses, sunfish, and darters appeared in 2000, the thirtieth anniversary of my first genetic analyses of cave fish and field mice back in graduate school at the University of Texas. I have come a long way since then, as has the field of molecular natural history. When I began such studies in 1970, the discipline (like me) was adolescent, swashbuckling, fresh, and untapped. Any empirical or conceptual topic at the interface of natural history and molecular biology was novel, and I could genetically explore almost any wild species, or any question about nature, without concern about bumping elbows with the very few other like-minded research groups at the time. Now, after three decades of sustained growth, the science of molecular ecology is sophisticated, mature, and crowded. I do sometimes miss the frontier spirit of those earlier years, but I still find this brand of research on nature's ways to be intensely rewarding.

17 Syntheses

Lasting contributions to the evolutionary or natural-history sciences often differ from those in other disciplines such as mathematics, physics, or molecular genetics in the following sense: they tend to be integrative works, rather than singular breakthroughs. Few specific studies in evolutionary biology or natural history have had the immediate transformational impact of, say, the splitting of the atom, or the discovery of DNA's double-helical structure. This probably explains why many physicists and mathematicians, for example, make their primary mark through some sensational discovery often at a rather young age, whereas most renowned evolutionary biologists and natural historians (from Charles Darwin to Theodosius Dobzhansky to modern workers like Ed Wilson) are remembered primarily for their lifelong syntheses.

Early in my career, I couldn't imagine writing a book-length treatise on any scientific topic. Producing a volume that would distill a broader discipline was far more daunting than churning out the run-of-the-mill research papers that are a necessary part of academic life. Yet many years later I felt compelled to attempt synthetic works. What changed was a growing sense of scientific competence borne of experience, coupled with a personal desire to unify particular disciplines and perhaps shape their direction.

In the 1960s protein-electrophoretic techniques sprang upon the fields of population biology and evolution. Next, methods for mtDNA analysis were invented and exploded in popularity during the 1970s, as did various DNA-fingerprinting procedures in the 1980s. During those early decades of the "molecular revolution," many prominent research labs in the United States and abroad became focused on one overriding issue: is the newly revealed molecular variation selectively neutral, or is it adaptive? In other words, are the protein-level and DNA-level polymorphisms mostly irrelevant to an organism's survival and reproduction, their maintenance in natural populations merely reflective of some balance between the addition of variation by neutral mutation and its subtraction by random genetic drift? Or, alternatively, does most molecular variation matter to an animal's or plant's well-being? The nascent field of molecular ecology became fixated on this one issue. It is only a slight exaggeration to state that more genetic laboratories were then devoted to studying possible adaptive variation at the alcohol dehydrogenase gene in fruit flies (a model system for the analysis of natural selection) than were involved in using molecular markers to study the ecology, behavior, and natural history of all other creatures combined.

This preoccupation with the possible adaptive significance of molecular variation was a logical outgrowth of the traditional Darwinian paradigm that ascribed a major role to natural selection in shaping the genes that underlie morphological and behavioral variation. But were garden-variety variations at the molecular level likewise sculpted mostly by selective demands, and if so, how? Three ensuing decades of arduous research by molecular evolutionists generally failed to give the kind of clear answers for which most researchers had hoped. Only after the fact was the enormity of this scientific undertaking adequately appreciated, and the futility of simplistic approaches fully realized. In a reflective commentary years later, Richard Lewontin, one of the field's founding fathers, wondered in print whether the introduction of allozyme methods had been a "milestone or millstone" for biology.

With my background in ecology and natural history, I had quite a different perspective on molecular variation. Almost regardless of the outcome of the thorny selection-neutrality debate, I saw vast potential in the use of genetic markers to unshroud numerous mysteries of the natural world. As illustrated in the preceding chapters, the genetic tags that nature herself applies to all creatures could be read to answer natural-history questions on topics ranging from the evolutionary origins of unisexual fishes to the reproductive biology of polyembryonic armadillos. Only during the 1990s and beyond did this research orientation, formerly a backwater in molecular genetics, truly come into its own. For example, the first scientific journal devoted to this emerging discipline, *Molecular Ecology,* did not appear until 1992.

At about that time, I determined to compose a textbook that would formally identify and give conceptual coherence to this underappreciated discipline. This effort was made possible by my first sabbatical. At most universities, faculty members are entitled to occasional leaves of absence, typically one semester off for each seven years of in-house service. The time normally is spent visiting other laboratories, gaining exposure to new ideas, learning new teaching or research skills, or otherwise enhancing one's academic skills. The University of Georgia has no formal sabbatical program, but similar arrangements can sometimes be made. In 1991, after sixteen years at the university, I won a Sloan Foundation Grant, which underwrote my writing effort. I spent the sabbatical year at the Hopkins Marine Station in Pacific Grove, California, accompanied by Joan and our daughter, Jennifer, then ten years old. The lab's director, Denny Powers, graciously provided office space.

The writing went well, and I completed the manuscript on schedule. *Molecular Markers, Natural History, and Evolution,* published in 1994, detailed how molecular variation unveiled in the laboratory could then be used to query nature about animal behavior, natural history, and phylogeny. Other treatises were available at the time on the general topic of molecular evolution, but they had focused on the molecules themselves as primary objects of scientific interest. My book was an attempt to engage "whole-animal" biologists with the field of molecular genetics and, conversely, to inform molecular biologists about the countless applications of their genetic methods in biodiversity arenas. Part of the challenge of bridging the studies of natural history and genetics was that many geneticists viewed natural-history studies (if they thought about them at all) as rather unsophisticated, and, conversely, many natural historians harbored resentment against the growing prestige of molecular genetics and its diversion of research funds from the time-honored organismal sciences. Where possible in the book, I strove to erode such hostilities and stereotypes.

The book filled a void and was an immediate success. It tapped a broad undercurrent of latent interest in genetic markers that previously had not been articulated in book-length form. The fields of molecular ecology and evolution have grown tremendously since then, and biology students today might find it hard to believe that the molecular and biodiversity disciplines had little contact until fairly recently.

Writing that book expanded my horizons and provided a new sense of accomplishment. Finally, I had secured some of the intellectual gains of my lifelong scientific labors. Ever since then, I have viewed normal research articles from my laboratory not as end products alone, but as stepping stones toward some broader framework. Like popular songs that rise in the charts for only a short time, even the best of scientific articles quickly fade from view. Synthetic treatments, on the other hand, are more like musical albums, thematic collections of tunes with potentially greater lasting value. The best-orchestrated of scientific treatises (not that mine necessarily qualify) are like symphonic masterpieces, with a wide sphere of influence.

The final chapter of *Molecular Markers* dealt with conservation issues and provided a springboard for *Conservation Genetics: Case Histories from Nature*. This volume, published in 1996, was coedited with my friend and colleague, Dr. Jim Hamrick. Based on our experience with natural plant and animal populations, Jim and I were familiar with the many ways that genetic data can inform conservation efforts. As detailed in earlier chapters, these include identifying distinctive population units for wildlife management, unmasking relevant aspects of animal or plant behavior and natural history, reconstructing phylogeographic histories of populations and species, and in general helping to describe the planet's genetic diversity, which is, after all, what conservation biology seeks ultimately to preserve.

Several chapters in my first two textbooks dealt with phylogeography, a discipline that I was to synthesize several years later in *Phylogeography: The History and Formation of Species*. Published in 2000, this treatise summarized at length how molecular-genetic data could document the genealogical component of geographic population structure (see chapter 12). Among the book's central themes was the nonequilibrium nature of the natural world, a view that is now well substantiated empirically, yet often departs from the traditional orientation of much of theoretical population genetics.

My other synthetic endeavor to date was in response to the often voiced concern that scientists should translate and communicate their fields to the general public. Years of lecturing to undergraduates had emboldened me to think that I might be able to convey scientific perspectives to a broader audience. There also was a subject of compelling societal relevance that I long had been anxious to explore: the intersection of evolutionary genetics with religion. Like an undaunted puppy with a huge bone, I began to chew on this unwieldy topic. After three years of library research (in the humanities as well as the science sections) and writing, what emerged is what I personally feel is my most creative and important work to date: *The Genetic Gods: Evolution and Belief in Human Affairs* (published in 1998).

The title alludes to the book's scientific thesis that genes (the "genetic gods") and the evolutionary forces that have sculpted them play many roles in human affairs traditionally reserved for supernatural deities. This genetic jurisdiction clearly includes influences over our bodily features at microscopic and macroscopic levels—our physiology, metabolism, and health—but also (to an arguable degree) to more ethereal aspects of human nature, including basic elements of our emotional dispositions, psychological profiles, personalities, and ethical predilections. Even the most sacrosanct of human affairs—sex, reproduction, aging, death, and proclivities for group and family allegiance—are phenomena of natural (rather than supernatural) origin, forged by real (rather than surreal) natural processes, and firmly embedded in our evolutionary-genetic heritage.

Genetic influences over human affairs often are indirect, mediated and modulated by diverse cultural exposures during development, and manifest through genetically based cognitive abilities unique to our species. Thus the book was *not* another treatment of the hackneyed "nature versus nurture" debate, which in any event is mostly a false dichotomy, because nearly all plant and animal traits emerge through the developmental interplay between genes and environment. Instead the book's primary goals were to document the extraordinary revolution in evolutionary-genetic understanding of human nature in the century and a half since Darwin and Mendel, and to contrast these scientific explanations with prior scenarios of supernatural causation that had characterized the best efforts of religion and philosophy across all prior millennia.

Science is not so much a body of knowledge as it is a principled manner by which phenomena are explored. The epistemological hallmarks of science are objectivity, critical open-mindedness, and the ruthless evaluation of alternative hypotheses against evidence. Although these ideals are not always met in practice, the methods of science contrast diametrically with traditional sources of knowledge such as divine revelation or unquestioned faith. In the last 150 years, applications of the scientific method in biology and medicine have revealed vastly more about the origins, makeup, and operations of life than had been learned in all prior centuries of human endeavor. Indeed, the objective findings of evolutionary genetics and the other life sciences have yielded virtually all that we scientifically know or currently surmise about the history and workings of the biological world.

Even in educated societies, most people remain ignorant of these awesome discoveries or hostile to their religious or philosophical implications. In the United States, for example, evolutionary biology seldom can be mentioned without unearthing deep antipathies. Many biology teachers are reluctant to even mention the topic of evolution for fear of backlash from parents, principals, and local school boards. How can this deplorable state of affairs have come to be? The historical reason, of course, is that following Charles Darwin, a logical scientific rationale no longer remained for invoking God's direct hand to account for life's diversity or for humans' biological position in the organic world. The apparent hegemony of evolutionary genetics in sacred realms was more than most fundamentalist religions could tolerate, objectivity and rationality be damned.

Yet, in some ways that matter, perhaps science and religion are not really so far apart. For example, in the centuries before Darwin, it was often the devoutly religious who most reveled in the wonder of nature, interpreting careful biological inquiry as a righteous means to celebrate the craftsmanship of God. In the ninth century, an Islamic movement within the Abbasis Empire was motivated by the notion that the surest route to the mind of God was through rational study of physics and the natural sciences. More recently, observant Christian naturalists of the sixteenth to nineteenth centuries ("natural theologians," such as William Bartram) were reverent in the traditional sense, seeking through their studies to glorify the works of a divine God.

Today evolutionary biologists similarly rejoice at the wonders of the natural world, though many of us might see the work of any purposeful deity, if such exists, as evidenced by mechanistic processes suitable for scientific inquiry within the physical rather than metaphysical realm. As phrased by the eminent late astronomer, Carl Sagan, perhaps "God is the sum total of all the laws of nature." A recent editorial in my local newspaper expressed it thusly: "Those who would celebrate the universe as creation must not flinch from it when seen clearly through the lens of science."

The Genetic Gods was an attempt to document current evolutionary-genetic thought and empirical evidence on human origins, health, behavior, and cultural practices, and also to make intelligible the ongoing revolution in gene-based technologies in human medicine. It was also an attempt to compare scientific understandings with those based on inspired faith and religious revelation. For example, I recounted delightful but fanciful Creation myths elaborated through the centuries by various cultures concerning the origins of life, of humankind, and of the individual in the womb, contrasting each of these stories with revelations of the evolutionary-genetic sciences.

My intention was not to be confrontational, but rather to explore possibilities for a more constructive accommodation between science and faith, between mind and heart, on topics of ecology and human health. Shouldn't self-confident religions welcome, even sponsor, scientific inquiries in evolutionary genetics and then savor the hard-won fruits of that labor as an invitation to enlightened deliberations on humankind's place within any broader scheme? Can we not find a way to harmonize religious and scientific appreciations of life in ways that would better serve humanity as we face the daunting environmental and medical challenges of the twenty-first century? Must science and religion continue to do battle even as the failure of Earth's overburdened life-support systems threatens to destroy us all?

My interest in the science/religion interface and its relevance to ecological issues probably traces to my childhood in nature at Ice Lake, coupled with lessons from Fountain Street Church in Grand Rapids, where I was introduced to what now might be called secular humanism. Upon completing *The Genetic*

Gods, I sent a copy to Fountain Street's minister, Duncan Littlefair, himself an author of several books on naturalism and religion. Although retired, Dr. Littlefair remained intellectually active, and he responded with a kind letter expressing gratitude for my efforts (and those of other scientists, like E. O. Wilson in *Consilience*) to try to draw open-minded religions closer to science. Yet he also couldn't help feeling a deep personal frustration in those regards, noting that religion in the United States, spurred by fundamentalist movements, generally is more retarded now than at any time in the last 100 years.

Part Four

Perspectives

At the start of the twenty-first century, native biotas are under assault worldwide. Coral reefs are dying, fisheries are collapsing, rain forests are disappearing, wetlands are being drained, exotic species are transplanted willy-nilly, extinction rates are escalating, and in general nature is retreating into smaller pockets as the continents fill with humans. Naturalists and evolutionary biologists are saddened by the realization that biodiversity, the subject matter of our inquiries, is in serious decline. Furthermore, we sometimes get discouraged by societies' inexcusably poor literacy on scientific and environmental issues. With such ignorance often comes misplaced hostility toward evolutionary biology, genetics, and ecology, ironically at a time when insights from these disciplines are most needed.

18 Reflections on Nature and Genetics

An academic career in biological research has been gratifying in ways that few other vocations could have matched. Where else but in academia could I have combined the emotional pleasures of natural-history study with the rigorous scholarship demanded by the evolutionary-genetic sciences? Where else could I have aspired through teaching and creative writing to instill an ecological and evolutionary awareness and an appreciation of the critical scientific method in so many young minds? Where else could I have found the freedom to pursue any line of inquiry, unfettered by top-down directives, conformity to societal norms, necessity for monetary profit, or any other external mandate? Where else could I have roamed library halls and forest paths with equal devotion? Where else could I have been a white-collar intellectual without going against my outdoorsman's blood?

In my field of study, numerous topics ranging from natural history and field ecology to molecular biology, mathematical modeling, and statistical analysis genuinely intersect. In the best of evolutionary-genetic research, all of these threads and others are woven into intelligible, durable fabrics. And what engrossing tapestries they can be, describing both the history and the operations of life itself! Where else do scientific discoveries speak so compellingly to the essence of nature's ways and thereby also to an objective view of humankind's place in any broader scheme? Where else but in the sphere of environmental issues is so much ultimately at stake? The biodiversity sciences (including evolutionary genetics) are at once among the most challenging, encompassing, and meaningful areas of human inquiry.

I want to close this book by paraphrasing from speeches that I recently delivered to graduating students and their parents at the Honors Day Convocation and the Christian Faculty Forum on the University of Georgia campus. My basic hope is that societies will somehow find a way to incorporate objective, scientific understandings of life into broader "religious" attitudes that may better serve humanity in the fullest sense. Only through such weddings is it likely that we will be able to confront the cultural and environmental challenges of the coming decades with some proper combination of intelligence *and* emotional commitment.

I told the audiences that we find ourselves now on a biological voyage of discovery, spearheaded by molecular and evolutionary genetics, that promises to challenge our most fundamental understandings of nature, ourselves, and indeed many of our most sacred philosophical, spiritual, and religious beliefs. If the twentieth century was the golden age of physics, marked by momentous achievements such as the birth of relativity theory, splitting the atom, and the beginning of space exploration, then the twenty-first century will surely be the golden age of biology.

Geneticists will soon decode the fully sequenced human genome, and medical researchers will plunge far more deeply into the brave new world of gene therapy and genetic engineering. Natural historians, geneticists, and systematists will complete the first full catalog of the planet's remaining flora and fauna, and reconstruct the sweeping phylogenetic tree of life. Biologists may soon probe even the most intimate workings of the human brain, of the cellular processes by which thoughts are generated and stored. In many respects, this coming century should be the best of times for biology.

In other respects, however, it may be the worst of times, likely to be remembered also as a period of global environmental deterioration and biotic extermination. Many experts believe that we are already in the midst of the greatest mass extinction event since sixty-five million years ago, when a large asteroid impacted the earth and extinguished the dinosaurs and many other life forms. Today, it is a single species that is impacting the planet. Through *Homo sapiens'* lack of intelligent self-restraint on population growth, we are simply overwhelming our planetary home, making life needlessly difficult, and crowding other species out of existence. We find ourselves now at a historical crossroads. Nothing less than the fate of the earth is at stake, ironically, even as the biological sciences develop extraordinary insights into the nature of life.

To place the current ecological situation in a useful historical context, I then briefly described to my audiences a delightful nonfiction work by a well-known natural historian of the late eighteenth century, William Bartram. In his travel diary, he described his six-year exploration, from 1773 to 1778, of a land filled with an ebullience of wildlife and plants. He wrote of a heavily forested land, with trees so large that from single trunks the natives built huge dugout canoes that could carry up to thirty warriors. He wrote of a land with abundant serpents, some ranging to ten feet long and thicker than a man's leg. He told of bright parakeets flying in flocks through the forests. He described pristine watercourses, clean and pure, so teeming with fish that unbaited hooks sufficed for their capture. He explained the perils of crossing rivers filled with crocodiles up to twenty feet long, sometimes so numerous that if they had kept still, he might have walked across their backs from shore to shore. He reported such a bounty of wild game that securing dinner at each evening's campsite presented little challenge, yet he also emphasized the necessity of hoisting leftovers into the trees lest they be disturbed by marauding tigers, bears, and wolves.

The exotic land that Bartram described was the southeastern corner of North America, as it existed a mere 200 years ago. (The "tigers," by the way, would now be called pumas, common at the time but now extinct in Georgia, as are Carolina parakeets, red wolves, and many other creatures that Bartram encountered). Huge trees of the sort that Bartram took for granted now occur only in a small parcel of North Carolina's Joyce Kilmer National Forest, essentially the last remaining speck of primary woodland in the southeastern United States. How quickly things have changed over these last two centuries! In a mere ten human generations, the Southeast, like most of the continent, has been extensively cleared, fenced, converted to agriculture and pasturage, fought over, paved, parceled for housing, and in general shaped and reshaped by human activities. If we in Georgia could somehow be magically transported back in time, I doubt that many of us would recognize Bartram's environs as our homeland.

With our hypothetical time machine, we might also contemplate going back further in time on our continent, for example, to 1000 A.D. Suppose we wanted a big night out on the town and chose to be transported to the grandest city in North America at that time. This was Cahokia, a large Indian settlement along the central Mississippi River, near what is now St. Louis. It was populated by about 20,000 residents, or about one-fifth as many people as reside in my present-day hometown, Athens, Georgia. Yet many modern urbanites would consider Athens a mere Podunk.

We might wish next to use our magical time machine to transport us back to a Georgian landscape as it existed ten to fifteen thousand years ago, just after the retreat of the last major continental glacier. This was at a time shortly after humans first arrived on the continent following their trek from Asia across the Bering land bridge. As good naturalists in the Bartram tradition, we would be impressed by the animals in this exotic realm: giant ground sloths, mastodons, glyptodonts (giant armadillo-like creatures), sabre-toothed tigers, dire wolves, and many other creatures that are no more.

From an evolutionary perspective, even 10,000 years is but a brief instant, an evening gone. Many abstract devices have been used to emphasize this point. For example, the fifteen-billion-year history of the universe might artificially be compressed into the more intelligible scale of one calendar year, beginning on January 1 and progressing to late in the day on December 31, one year later. On this condensed scale, life on earth arose on about October 9. By December 17, the first multicellular invertebrate animals began to flourish, and the first vertebrates appeared on December 19. The dinosaurs arose and persisted for more than four days, from December 24 to 28. Humans appeared at 10:30 P.M. on the final day of the yearly calendar, December 31. The birth of Christ occurred an instant before midnight, at 11:59 P.M. plus 56 seconds. The scientific method has been widely employed only during the last second on this year-long cosmic scale.

It is also within this last second of time that the global human population has risen so rapidly in size, from only a few hundred million people prior to the industrial revolution to more than six billion today. Projections for the current century are that ten to twenty billion people will soon crowd the earth, all trying to wrestle from it a decent standard of life. Astonishingly, our species now shows a net increase of more than a quarter million people per day. As you read this chapter, more than 10,000 individuals will be added to the earth's burgeoning human population, nearly enough to populate Cahokia, the largest city in North America just 1,000 years ago. As an ecologist and natural historian who cherishes biodiversity, I find these facts deeply troubling.

On the other hand, as an evolutionary geneticist, I can't help looking forward to the twenty-first century with excitement and hope. Especially intriguing at the moment is the Human Genome Sequencing Project. In the year 2000, the first human chromosome, our genome's second smallest, was fully sequenced. Chromosome 22 proved to contain about 545 operational genes, mutations at some of which are implicated in several serious human genetic disorders. Before too long, the function of all 30,000 human genes scattered across our twenty-three pairs of chromosomes will be elucidated.

In medicine, the implications of such genetic discoveries are nothing short of astounding, and they soon will lead us far beyond mere description and diagnosis alone into the bold new realm of gene manipulation. Just a few centuries before the birth of Christ, Hippocrates argued that the workings of the human body are natural rather than supernatural. Now, 2,500 years later, for better or maybe sometimes for worse, we stand on the threshold of molecular technologies that enable us to tinker directly with our own genes and those of other animals and plants.

In the history of medicine and health care, there have been at least three major revolutions. The first occurred in 1854, after the British surgeon John Snow discovered that cholera is spread by contaminated water. This led to public sanitation measures that helped to curb infectious diseases. The second revolution, also in the mid-1800s, involved the development of surgical procedures under anesthesia. These enabled doctors to fix physical ailments such as appendicitis. A third revolution involved the introduction of antibiotics and vaccines, which helped to prevent or cure many infectious diseases.

Many prognosticators believe that we are now at the doorstep of the fourth medical revolution—human genetic engineering. In the not-too-distant future, gene therapy may well become standard practice in treating genetic diseases such as sickle cell anemia, Huntington disease, various cancers, heart disease, and a vast array of other inborn errors of metabolism. Such applications may lead our grandchildren to marvel at their grandparents' fortitude in the dark ages before gene therapy, much as we may marvel at our own great-grandparents' fortitude in facing life's hardships without benefit of anesthesia or refined surgical procedures.

From a philosophical perspective, perhaps of even greater significance are molecular findings that challenge theistic claims of omnipotent perfection in life's operational design. The human genome upon close inspection appears to be a legacy-laden patchwork of DNA. For example, scattered along the length of chromosome 22 are 134 "pseudogenes," dead genes no longer functional. And, an astonishing 42 percent of chromosome 22 consists of repeated DNA elements which proliferate somewhat like viruses, suggesting that they have selfish rather than cell-serving motives. These and many other details of molecular machinery give every indication of historical contingency, evolutionary tinkering, and genetic improvisation by nature.

We may soon anticipate improved scientific understanding of genetic issues at other levels as well, such as why the process of aging (as opposed to individual immortality) and of sexual (as opposed to asexual) reproduction are nearly ubiquitous to life. The broader point is that the field of evolutionary genetics, with roots little more than a century deep, will continue to offer objective insights into human conditions that until recently had been the exclusive purview of philosophy and religion.

In the allied fields of ecology and environmental resources, scientific knowledge is likewise increasing at a remarkable pace. Phenomena such as global warming, the biotic effects of industrial pollutants, and global declines in biodiversity were not even recognized as important issues just fifty years ago, but now they are standard topics of scientific investigation, as are (in some cases) potential technological or other solutions. Especially in the last twenty-five years, the environmental sciences have far outstripped societal capacities to incorporate and translate the new understandings into wise policy.

What will the future hold? Of course no one knows, but it is neither hubris nor the standard myopia of contemporary existence which leads me to believe that the twenty-first century will be uniquely significant in the history of life on earth. It is literally true that at no other point in life's grand expanse of four billion years have the actions of one species so dictated the fates of all others. In this decisive time, human hands collectively clutch the planet's jugular vein. Will we strangle the earth and thereby ourselves, or will we discover the means to convert the grip to a far gentler caress?

Perhaps this century will witness a sympathetic awakening of philosophy and religion to the discoveries of the natural sciences and a translation of these into constructive societal attitudes with regard to environmental challenges. Perhaps we can yet envision some new merger of the qualities of wonder and reverence for our surroundings that religion can provide, with the equally inspirational understanding of ourselves and of the natural world that can come from rational scientific appraisal.

I have sometimes concluded my campus talks by noting that when William Bartram roamed the southeastern forests 200 years ago, he sought through his naturalist studies to glorify the magnificent works of God. On nearly every page of his diary, Bartram expressed a sense of wonder:

> This world, as a glorious apartment of the boundless palace of the sovereign Creator, is furnished with an infinite variety of animated scenes, inexpressibly beautiful and pleasing, equally free to the inspection and enjoyment of all his creatures.

When Darwin set out on his voyage of discovery aboard the HMS *Beagle* forty years later, he too was a natural theologian seeking to understand the nature of Creation. He had no idea that his later scientific synthesis would so revolutionize rational thought about nature and human's place within it. Yet, like Bartram, he retained a sense of the magnificence of it all, as illustrated by his closing paragraph in *The Origin of Species:*

> It is interesting to contemplate a tangled bank, clothed with plants of many kinds, with birds singing on the bushes, with various insects flitting about, and with worms crawling through the damp earth, and to reflect that these elaborately constructed forms, so different from each other, and dependent upon each other in so complex a manner,

> have all been produced by laws acting around us. . . . There is grandeur in this view of life, with its several powers, having been originally breathed by the Creator into a few forms or into one; and that, whilst this planet has gone cycling on according to the fixed law of gravity, from so simple a beginning endless forms most beautiful and most wonderful have been, and are being evolved.

Open-minded religions need not be at odds with the ecological and evolutionary-genetic sciences. As we move forward into the twenty-first century, a difficult and pressing challenge will be to achieve a useful consilience between alternative ways of "knowing"—an integration of mind and heart, of analytical thought and emotional dedication, especially with regard to environmental and related social issues. Perhaps we can rise above our fears and ignorance, and exorcise the demons that obstruct our collective vision on how to achieve a sustainable relationship with nature. Perhaps we can somehow assimilate our growing knowledge from the life sciences into philosophical outlooks and responsible modes of action that will reflect a wiser appreciation of natural history, of our genetic heritage, and, ultimately, of ourselves.

Index

L

M

N

O

P